THE S-CALM MODEL
THE APPLICATION OF ETHICAL LEADERSHIP IN THE MILITARY

Dedicated to my family, whose love has sustained my ethical leadership throughout my career.

The S-CALM Model
The Application of Ethical Leadership in the Military

DR DENNIS VINCENT MBE

Howgate Publishing Limited

Copyright © 2025 Dennis Vincent

First published in 2025 by
Howgate Publishing Limited
Station House
50 North Street
Havant
Hampshire
PO9 1QU
Email: info@howgatepublishing.com
Web: www.howgatepublishing.com

All rights reserved.

No part of this publication may be reproduced, stored in a retrieval system, or transmitted in any form or by any means including photocopying, electronic, mechanical, recording or otherwise, without the prior permission of the rights holders, application for which must be made to the publisher.

British Library Cataloguing-in-Publication Data
A catalogue record for this book is available from the British Library

ISBN 978-1-912440-71-9 (pbk)
ISBN 978-1-912440-72-6 (hbk)
ISBN 978-1-912440-73-3 (ebk – ePUB)

Dennis Vincent has asserted his right under the Copyright, Designs and Patents Act, 1988, to be identified as the author of this work.

The views expressed in this publication are those of the author and do not necessarily reflect official policy or position.

CONTENTS

Tables viii
Figures ix
Preface x
Acknowledgements xii
Abbreviations xiii

1. Ethical Leadership and Background Concepts — 1
2. Situational Influencers — 17
3. Common Individual Behaviours — 35
4. Common Group Behaviours — 54
5. The Flow of Unethical Actions — 71
6. S-CALM: S – Situational Influencers — 85
7. S-CALM: C – Common Behaviour (Individual) — 102
8. S-CALM: C – Common Behaviour (Group) — 122
9. S-CALM: A – Accountability — 137
10. S-CALM: L – Leadership — 153
11. S-CALM: M – Moral Compass — 170
12. Application of the S-CALM Model — 185

Appendix 1: Decision Making Questionnaire 1 196
Appendix 2: Ethical Decision Making Coaching Tool 199
Appendix 3: Decision Making Questionnaire 2 205
Bibliography 208
Index 224

TABLES

Table 1.1	Ethical Leadership Pillars	3
Table 7.1	Barriers to Change	116
Table 8.1	Common Individual and Group Behaviours	136
Table 9.1	Types of Accountability	140
Table 10.1	Offering and Receiving a Challenge	167
Table 11.1	British Army Core Values Spectrum	177
Table 11.2	British Army Values and German Army Virtues	178

FIGURES

Figure 1.1	Components of the S-CALM Model	7
Figure 1.2	The S-CALM Model	7
Figure 2.1	Situational Influencers	19
Figure 2.2	Overlap of Situational Influencers	19
Figure 5.1	Flow of Unethical Actions	72
Figure 5.2	The Flow of Unethical Actions – Sergeant Blackman Case Study	83
Figure 7.1	Situation and Personality	118
Figure 12.1	Application of the S-CALM Model	187
Figure 12.2	Ethical Hierarchy	194

PREFACE

The genesis of this book was the requirement to provide a suitable ethical leadership text for the trainee officers at the Royal Military Academy Sandhurst, therefore there is a slight focus on the British Army. However, I have tried to balance this by benchmarking the doctrine of the US Armed Forces and the Australian Defence Force as they have recently developed military ethics doctrine. I have also added points from other nations that are investigating military ethics such as Canada, the Netherlands and German.

The book is a combination of my lived experience as a British Army officer for nearly 30 years, my research into military ethics over the last 12 years and my teaching of ethical leadership at the Royal Military Academy Sandhurst for 9 years. I have started each chapter with a personal story from my military career which refers to the subject covered in that chapter. For my initial research into the S-CALM model, I focused on the 20 case studies that have been used for military ethics education at the Royal Military Academy Sandhurst, but I added additional less well-known case studies as the research developed. Reviewing these case studies gave me the data for establishing the situational influencers and common behaviours that occur in most situations. However, the hardest part of the research was finding positive case studies which demonstrate how these influencers and behaviours can be mitigated. I was lucky to have a friend in Richard Westly who shared his experience in Gorazde to reveal how a leader can recognise and mitigate the situational influencers. Nevertheless, finding constructive case studies for the common behaviours was more challenging as people tend not to record these positive actions as well as negative ones.

This book also includes my research into the effectiveness of the S-CALM model, which demonstrates its strength in preparing trainee officers for future, complex operating environments. The research also established some issues in the way that the Royal Military Academy Sandhurst taught military ethics and this in turn led to the development of the Ethical Decision Making Coaching Tool to rectify these shortfalls.

I hope that you enjoy this book and find the S-CALM model and the STOP Protocol useful tools in making good, ethical decisions under the power of the situation.

Dr Dennis Vincent MBE

ACKNOWLEDGEMENTS

I would like to thank the previous Commandant, Major General Zac Stenning, and the previous Deputy Commandant Officer Training, Colonel Mark Gidlow-Jackson, of the Royal Military Academy Sandhurst for allowing the ethical leadership research to be conducted in 2023, which assisted in validating the model. Thanks also need to go to the teaching team at the Department of Communication and Applied Behavioural Science, who assisted in the refinement of the model by teaching it and offering constructive criticism and advice. I would also like to thank Colonel (Retired) Richard Westley OBE MC who shared his experiences in Gorazde with me and allowed me to use our interview to highlight how he resisted the power of the situation whilst in command. Finally, I would like to thank Major Ben Ordiway formally an instructor at the US Military Academy West Point, for his collaboration on the design of the Ethical Decision Making Coaching Tool.

ABBREVIATIONS

AAR	After Action Review
ADF	Australian Defence Force
BG	Brigadier General
BLAME	Behaviours, Leadership, Accountability, Moral Compass and Effects of the Situation
BSA	Bosnian Serb Army
CABS	Department of Communication and Applied Behavioural Science
CAL	Centre for Army Leadership
CO	Commanding Officer
HMS	Her Majesty's Ship
ICRC	The International Committee of the Red Cross
IED	Improvised Explosive Device
KFOR	Kosovo Force
LOAC	Law of Armed Conflict
LAP	Lawful, Acceptable and Professional
Lt Col	Lieutenant Colonel (UK)
LCpl	Lance Corporal
LTC	Lieutenant Colonel (US)
LTG	Lieutenant General
Maj	Major
MOD	UK Ministry of Defence
MP	Military Police
NATO	North Atlantic Treaty Organisation
NCO	Non Commissioned Officer
NICRA	Northern Ireland Civil Rights Association
OCdts	Officer Cadets
OP	Observation Post
PARA	The Parachute Regiment
Pte	Private

RMAS	The Royal Military Academy Sandhurst
ROE	Rules of Engagement
RSM	Regimental Sergeant Major
SAS	Special Air Service

1
Ethical Leadership and Background Concepts

Introduction

As a platoon commander, one of my young soldiers was killed off duty by the Provisional Irish Republican Army (PIRA) in Northern Ireland. Shortly after his death, my platoon began four weeks of patrolling duties in the heartland of the PIRA. The platoon wanted revenge for the death of one of their comrades. However, I knew that this was the wrong thing to do and had to demonstrate strong ethical leadership to ensure that they did not enact reprisals on the local population. I was aware that one of my sister platoons in the company had begun to throw paint bombs at Republican murals. We lived in a couple of portacabins in the car park of a police station and on an inspection of our platoon base, I found some milk bottles filled with paint hidden in the drains; I destroyed them. Having done so, I gathered my platoon together and reminded my soldiers why we were deployed in Northern Ireland and the importance of legitimate actions. I also informed them that I would take disciplinary action on anyone found with a paint bomb. This was a difficult time, ethically I stood alone as the platoon commander and I too was hurting at the death of one of my own men, however I was able to lead my platoon through this difficult time. I believe that they subsequently understood the importance of retaining the moral high ground and not sinking to the level of the terrorists.

I had a number of similar experiences throughout my 30-year career in the British Army and when I became an academic, I wanted to reflect and understand how most leaders make sound, ethical decisions, while others make poor decisions and conduct unethical actions. This book is based on my research and teaching of ethical decision making in the Department of Communication and Behavioural Science (CABS) at the Royal Military Academy Sandhurst (RMAS). This first chapter will explore what ethical leadership is and how it is approached in Western armies. It will then outline my ethical decision-making toolkit, known as the S-CALM model, which is currently taught to all British Army officer trainees at RMAS. It will move on to a explore a little about the use of case studies as a way of understanding ethical dilemmas. It will then study background concepts, consisting of research conducted in the 1960s and 1970s that were the starting point for the S-CALM model. The chapter will conclude with a guide to the subsequent chapters.

What is Ethical Leadership

Ethical leadership is defined as 'the demonstration of normatively appropriate conduct through personal actions and interpersonal relationships, and the promotion of such conduct to followers through two-way communication, reinforcement and decision-making.'[1] As an academic theory ethical leadership theory has been discussed since around 2005. Linda Trevino, Lura Hartman and Michael Brown were early scholars in it and they demonstrated that it is a cross-cultural phenomenon and has global reach; their research showing that although there were slight regional variations 'the content of ethical leadership appears to be universal.'[2] In practical terms, they believed that ethical leadership consists of 'two behavioural components: ethical person and ethical manager.'[3] Table 1.1 below shows what they refer to as the two pillars of ethical leadership.

The ethical person component of ethical leadership is concerned with the moral character of the leader. Trevino, Hartman and Brown's research indicated that 'being an ethical person is the substantive basis of ethical leadership' however this was caveated by them stating that a person's 'ethical core must be authentic' otherwise it will not have impact.[4] They went further to state that a leader who was not perceived to be a 'strong ethical person

---

[1] Brown et al, 'Ethical Leadership: A social learning perspective for construct development and testing', *Organizational Behavior and Human Decision Processes*, Vol. 97 (2005), 120.
[2] Michael Brown and Linda Trevino, 'Ethical Leadership: A Review and Future Directions', *The Leadership Quarterly*, Vol. 17 (2006), 613.
[3] Sehrish Ilyas, Ghulam Abid and Fouzia Ashfaq, 'Ethical Leadership in sustainable Organizations: The Moderating Role of General Self-efficacy and the Mediating Role of Organizational Trust', *Sustainable Production and Consumption*, Vol. 22 (2020), 196.
[4] Ilyas, Abid and Ashfaq, 'Ethical Leadership in sustainable Organizations', 130.

Ethical Leadership	
Ethical Person	Ethical Manager
Personal Traits Personal Behaviours Decision-Making	Role Model Through Visible Actions Communicating Ethics and Values Rewards and Discipline

Table 1.1 Ethical Leadership Pillars

but who attempts to put ethics and values at the forefront of their leadership agenda is likely to be perceived as a hypocritical leader.'[5] They believed that an ethical person demonstrated their authenticity by their traits, behaviours and decision-making.

The traits expected of an ethical leader revolved around the principles of trust and fairness. They also considered the concepts of honesty, trustworthiness and integrity important, stating that most studies showed that ethical leaders were 'thought to be honest and trustworthy.'[6] The traditional behaviours of an ethical leader were not the cognitive behaviours that will be explained later in this book, but were more aligned with virtue ethics. An ethical leader was one described as a person that can 'do the right thing' in difficult circumstances, display 'concern for people' and has a strong 'personal morality.'[7] When making decisions, ethical leaders were seen 'as fair and principled decision-makers who care about people.'[8] They were thought to 'hold to a solid set of ethical values and principles' and the decisions they made were considered to be fair and objective and 'follow ethical decision rules.'[9] Therefore the ethical person was one who was intrinsically moral and naturally made good ethical decisions.

The ethical manager component of the theory is about the leader's conduct which is aimed at encouraging the ethical behaviour of followers. The ethical leader is the exemplar of moral actions to their followers. The research revealed that 'ethical role models showed care, concern and compassion for people' and the leader held 'themselves to high ethical

---

5 Ibid., 138.
6 Brown and Trevino, 'Ethical Leadership: A Review and Future Directions', 597.
7 Linda Trevino, Lura Hartman and Michael Brown, 'Moral Person and Moral Manager: How Executives Develop a Reputation for Ethical Leadership', *California Management Review*, Vol. 42, No. 4 (2000), 131.
8 Brown and Trevino, 'Ethical Leadership: A Review and Future Directions', 597.
9 Ibid.

standards.'[10] An important aspect of managing in an ethical way was seen as the need to for a leader to communicate a consistent ethical vision to their team. In most organisations this was done by the socialisation of a set of core values or virtues that followers were expected to adhere to. Finally, an ethical manager was seen as one that would apply 'rewards [for] ethical conduct and disciplines unethical conduct.'[11] This requirement to apply extrinsic motivation to followers was considered a key one for changing teams' actions. Therefore, the application of a transactional leadership style was recognised as an imperative if an ethical leader was to maintain high ethical standards in their organisation and hold those in their team accountable for their unethical actions, whilst rewarding those acting in an ethical manner.

The Trevino, Hartman and Brown idea 'affords similarity with transformational and authentic leadership.'[12] There are four components of authentic leadership: self-awareness, internalised moral perspective, balanced processing and relational transparency. There are also four components of transformational leadership: individualised consideration, intellectual stimulation, inspirational motivation and idealised influence. There is little doubt that some of these components overlap with the concept of ethical leadership as explained. However, neither of these other styles of leadership has the crucial aspect of the application of rewards and punishment in order to motivate followers and it is this 'transactional influence process that distinguishes' it from the other modern leadership styles.[13] The theory believes that for an ethical leader to be successful they must be able to apply this extrinsic motivation to their teams to ensure that they act in an ethical manner. The British Army agrees that this application of rewards for good ethical behaviour 'provides incentive and reinforces what 'right' looks like, building self-esteem and confidence.'[14]

What is Ethical Leadership like in the Army

Ethical leadership in the British Army is about making an ethical decision under stress, rather than a separate theory of leadership. An ethical decision is defined as 'a decision that is both legal and morally acceptable to the larger

---

10 Gary Weaver, Linda Trevino and Bradley Agle, 'Somebody I Look Up To: Ethical Role Models in Organisations', *Organizational Dynamics*, Vol. 34, No. 4 (2005), 318.
11 Trevino, Hartman and Brown, 'Moral Person and Moral Manager', 136.
12 Ilyas, Abid and Ashfaq, 'Ethical Leadership in sustainable Organizations', 196.
13 Brown and Trevino, 'Ethical Leadership: A Review and Future Directions', 599.
14 MOD, *Army Leadership Doctrine*. (London: HMSO, 2021), 89.

community.'¹⁵ The British Army's Centre of Army Leadership defines it as being 'leadership that is directed by respect for ethical beliefs and values and for the dignity and rights of others.'¹⁶ It goes on to state that ethical leadership is 'related to concepts such as trust, honesty, consideration and fairness.'¹⁷ The US Army has no clear definition for ethical leadership, but does discuss ethical reasoning, in which its states that 'to be an ethical leader requires more than merely knowing the Army Values. Leaders must be able to live by them to find moral solutions to diverse problems.'¹⁸ The Australian Defence Force also has no clear definition but in its ethical doctrine it has the following statement 'ethical leadership is the single most important factor in ensuring the legitimacy of our operations and the support of the Australian people.'¹⁹ The North Atlantic Treaty Organisation (NATO) describes ethical leadership as requiring 'understanding and critical thinking skills that must be cultivated and exercised throughout a leader's development.'²⁰

This NATO approach which includes the requirement to use behavioural science theories in ethical thinking goes against most Western armies' ideas which are mostly focused on virtue ethics concepts. However, as early as 2011 the eminent military ethicist, Professor David Whetham recognised that 'V & S [Values and Standards] training cannot on its own give enough guidance to cover what ought properly to be considered as operational ethics practice.'²¹ A 2014 survey of staff and senior trainees at RMAS asked what people believed that an ethical style of leadership looked like, many believed that it was 'the cultural embodiment of 200 years of training', however the respondents felt that the style itself was not well defined.²² Also in 2014, a report on the Sergeant Blackman murder of a Taliban fighter in Afghanistan stated that adhering to the core values would not equip leaders in the future

---

15 Thomas, Jones, 'Ethical Decision-Making by Individuals in Organizations – An Issue-Contingent Model', *Academy of Management Review*, Vol. 16 (1991), 367.
16 Centre of Army Leadership, *Army Leadership Doctrine: What Leaders Are Workshop*. (Camberley: CAL, 2019), Slide 12.
17 Ibid.
18 US Army, *ADP 6-22: Army Leadership and The Profession*. (Department of the Army, 2019), 2-6.
19 ADF, *Leadership*. (Australia: Directorate of Information, 2021), 3.
20 NATO, *Leader Development for NATO Multinational Military Operations, STO-TR-HFM-286*. (Boston Spa: NATO, 2022), 9-2.
21 David Whetham, 'Ethics Education and Training' in *Defence Academy Ethics Seminar: To consider the Ethical Component of Military Capability*, ed. Patrick Mileham (London: Royal College of Defence Studies, 2011), 11.
22 Dennis Vincent, *Be, Know or do? An analysis of the Optimal Balance of the Be, Know, Do Leadership Framework in future Training at the Royal Military Academy Sandhurst*. Sandhurst Occasional Paper No. 20. (Sandhurst: Central Library, 2015), 22.

as 'contemporary and future operating environments are likely to pose highly complex problems.'[23] In the same year RMAS held a leadership conference to discuss a number of issues. One of the conference sessions discussed ethical leadership and at this it was the 'consensus was that while V & S [Values and Standards] in combination with LOAC [Law of Armed Conflict] was absolutely essential, they are not sufficient in equipping officers and soldiers with the necessary "toolbox" with which to deal with "real world" ethical problems in complex operational environments.'[24]

CABS at RMAS has been discussing ethical decision making with students since the introduction of the British Army Values and Standards, but the work had little form. To fill the requirement for a more developed ethical toolbox, research began in 2015 on how this information could be better presented and by 2017 the BLAME Model was developed and taught at RMAS to all British and international trainees. BLAME stood for: Behaviours, Leadership, Accountability, Moral Compass and Effects of the Situation. BLAME was used as a way of remembering the various parts of what the research had identified as forming unethical action and it also offered some limited ways of dealing with the effects. There were two catalysts to turning the BLAME idea into the S-CALM model, the first was the Coronavirus pandemic, which gave time for further detailed research. The second was a report by the British Army Inspectorate which fully endorsed the concept of BLAME but did not like the word blame and asked for the model to be repackaged. Therefore, in 2022, after further detailed research the design of the toolbox was improved and amended into the S-CALM model which is now taught to all trainees. During 2023 a research survey of Officer Cadets on the Regular Commissioning Course at Sandhurst was conducted to establish the usefulness of the model and its teaching. The results from this survey are included throughout this book. Prior to reading further, you might wish to take the survey, which is at Appendix 1.

Outline of The S-CALM Model

The S-CALM model is currently taught to all officer trainees at RMAS and to selected units in the British Army. It is a simple toolkit for remembering the key principles of ethical decision making in times of heightened stress

---

23 Navy Command, Telemeter-Internal Review, 7 March 2014, C3.
24 Stephen Deakin, *Leadership: Proceedings of a Symposium Held at the Royal Military Academy Sandhurst, April 2014. Sandhurst Occasional Paper No. 18.* (Sandhurst: Central Library, 2014), 33.

Ethical Leadership and Background Concepts

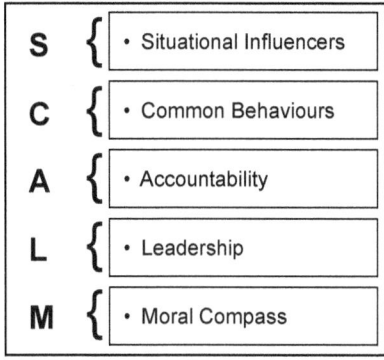

Figure 1.1 Components of the S-CALM Model

Figure 1.2 The S-CALM Model

especially when in command in a difficult situation. It has been designed to encompass ideas from four disciplines that are often not put together, which is what gives it use as an applied tool. The four components that combine to create the S-CALM model are: ethical leadership theory, military ethics, military leadership doctrine and behavioural science concepts as shown in Figure 1.1.

These four components are brought together in the model in a straightforward, understandable format which acts as a handrail for leaders cognitive processes to aid their ethical decision-making, especially when they are faced with the fatigue, stresses and emotions of command. The model can be recalled using the simple acronym S-CALM. This stands for: Situational Influencers, Common Behaviours, Accountability, Leadership and Moral Compass as seen in Figure 1.2. The detail of each part will be explained in chapters 6 to 11.

The model is applied by using another simple procedure known as the STOP Protocol. This is explained in more detail in chapter 11, but in outline it stands for:

S – Stop what you are doing. If actions are perceived to be unethical they should not be continued.
T – Take a few deep breaths. This enables rational thinking to engage.
O – Observe any situational influencers and common behaviours.
P – Proceed considering accountability, leadership and moral compass.

Case Studies

Case studies have been used extensively in both the research and explanation of the S-CALM model. Research has shown that 'a good moral education addresses both the cognitive and affective dimensions of human nature. Stories are an irreplaceable medium for this kind of moral education.'[25] In addition, research has demonstrated that understanding a situation by exploring past experience, especially in ethical education, by 'using case studies of war crimes involving military personnel from the same service and country as themselves. This would help military personnel realize that people just like them can become torturers, can kill civilians, and so forth.'[26] Therefore, case studies were chosen as the primary model for this research to deal with why and how unethical actions happened. The academic Robert Yin maintains that the 'how and why questions are more explanatory and likely to lead to the use of case studies.'[27] Whilst J. Gerring expands this to describe that the case study is 'an intensive study of a single unit for the purpose of understanding a larger class of (similar) units.'[28] In this book the unit to be studied is the S-CALM model for the purpose of understanding how unethical actions happen. However, this research is not a simple, single case study, but what Yin terms an 'embedded case study.'[29] An embedded case study focuses on a single unit, the S-CALM model, but attention is also given to sub-units, the various case studies. Therefore, readers of this book not only get an understanding of the S-CALM model but can also analyse past case studies and comprehend how they came about and what could have been done to stop them from taking place.

Background Concepts

Before getting into an understanding of how the research for this book was conducted and the details of the S-CALM model are explored, the early examination of unethical actions should be examined. What I call the

---

25 Vigen Guroian, *Tending the Heart of Virtue*. (Oxford: Oxford University Press, 1998), 20.
26 Matthew Talbert and Jessica Wolfendale, *War Crimes: Causes, Excuses and Blame*. (Oxford: Oxford University Press, 2019), 152.
27 Robert Yin, *Case Study Research: Design and Methods*, 2nd ed. (London: Sage Publications, 1994), 6.
28 J. Gerring, 'What is a case study and what is it good for?', *American Political Science Review*, Vol. 98, No. 2 (2004), 342.
29 Yin, *Case Study Research*, 41.

'Background Concepts' to the S-CALM model came from research conducted primarily in the 1960s following investigation into the activities of the Nazis. The three background concepts are: the banality of evil, incremental steps and the power of the situation.

Banality of Evil

Post the Second World War, most people wanted to believe that the Nazis' were monsters or psychopaths. The classic image of an SS officer is the psychopathic Commandant of Plaszow Camp, Amon Goeth, who was brought to life in the book *Schindler's List*. However the idea of the ordinariness of the action conducted by the Nazis' came sharply into focus in 1960 when the Israeli Mossad captured the Nazi Adolf Eichmann in South America. Eichmann had been one of the architects of what the Nazi's termed the final solution. He was a logistician and was principally responsible for the movement of five million people into the extermination camps and their efficient murder and disposal of the bodies. After his arrest he was taken to Jerusalem for trial. Hannah Arendt, a reporter from the *New Yorker*, was covering the trial. Arendt commented that 'the trouble with Eichmann was precisely that so many were like him, and that the many were neither perverted nor sadistic, that they were, and still are, terrifyingly normal.'[30] Arendt said that the problem was death was hidden behind bureaucratic blandness and indifference. Her perception was confirmed when Eichmann was examined by six psychiatrists prior to the trial the report 'ascribed to Eichmann sensitivity, talent, and spontaneous empathy.'[31] Over the course of the 14 week trial, she concluded that most evil is done by people who never make up their minds to do good or evil. From her experience, Arendt coined the term 'The Banality of Evil'. If something is banal, it is normal, everyday. She believed that anyone could be an Eichmann. The central theme of the 'Banality of Evil' is that everyone has the ability to do evil acts, given the right situation. George Mastroianni conducted research into why people commit unethical action, in this he concluded that 'most of the people who commit

---

[30] Hannah Arendt, *Eichmann in Jerusalem: A Report on the Banality of Evil*. (New York: Viking Press, 1963), 276.
[31] Brunner, Jose, 'Eichmann's Mind: Psychological, Philosophical, and Legal Perspectives', *Theoretical Inquiries in Law*, Vol. 1 No. 2 (2008), 7.

ethical transgressions in wartime are largely indistinguishable from soldiers who do not commit such offenses.'[32]

Incremental Steps and the Spiral of Violence

In 1961 Yale University psychologist Stanley Milgram was intrigued by the Eichmann trial and wanted to establish if Americans could have conducted crimes like the Holocaust. Over three months he conducted a series of experiments to discover how ordinary Americans would react to an authority figure. This book will only outline the results of the main experiment. In this main experiment, there were three people involved: a teacher who was an ordinary member of the public, a learner who was an actor and an authority figure who was a Yale employee. The learner was put behind a screen and attached to electrodes. The teacher was told to give a 15-volt electric shock to the learner if they answered a memory question wrong. The shocks, which were not real, were increased by 15 volts for every question that the learner got wrong. The teacher had a row of switches from the 15 volt one labelled 'slight shock' through 300 volts which was labelled 'danger' to 450 volts which was simply labelled 'XXX'. At 330 volts you would become unconscious and at 450 volts you would die. The Yale scientists anticipated that less than 1 per cent of teachers would go over 150 volts and that only 0.1 per cent would deliver the full range of shocks. They were surprised when all participants gave over 300 volts before stopping and 65 per cent of teachers delivered the full 450 volts.

There were many lessons drawn from Milgram's experiment, however the idea that evil starts with small steps is an important one. Once the teachers initially accepted shocking the learner with very small voltage, the majority quickly administered a lethal amount. Robert Cialdini is a psychologist who has investigated how people are persuaded and influenced to do things. One of the principal influencers is to appeal to peoples' need to be consistent in their actions. Therefore, once the teachers had committed to giving the first shock, the social obligation to be consistent meant that they were more likely to give subsequent shocks. This need for people to be consistent in their actions should not be underestimated. Dr Phillip Zimbardo confirmed that 'starting the path toward the ultimate evil act with a small, seemingly insignificant first step, the easy "foot in the door"' can lead to unethical

---

[32] Mastroianni, George, 'The Person-Situation Debate: Implications for Military Leadership and Civilian-Military Relations', *Journal of Military Ethics*, Vol. 10, No.1 (2011), 7.

actions.'³³ Therefore people can quickly move from doing small acts to larger acts and it has been shown that 'serious crimes often have small beginnings, and people refer to a "slippery moral slope".'³⁴ The Australian Defence Force has identified this issue and in its *Military Ethics* doctrine:

> When people repeatedly behave in a manner that deviates from professional standards without negative consequences, it can encourage and reinforce thebelief that the behaviour is justified and acceptable. Often such behaviour has its genesis in small changes from standard procedures that, over time, become significant breeches of professional standards.³⁵

The International Committee of the Red Cross (ICRC) has also noted that unethical actions tend to start with these incremental steps and refer to the process as 'a spiral of violence in which violations may become not only more and more serious but also more and more acceptable in the eyes of those who commit them.'³⁶ Each time that an individual or group commits an act there is a strong obligation to retain consistency with this action and continue. There then comes a point when to stop would mean admitting that the actions already conducted were unethical, therefore people commit to further acts and actions spiral out of control. The vital point for leaders is to stop these small acts early, before they spiral out of control.

The Power of the Situation

Milgram alluded to the power of the situation that causes people to act in ways against their rational thinking. However, most people do not recognise how their environment influences them. Therefore, they would tend not to agree that the 'behaviour that we think is under control of central motives, or traits, is really very sensitive to situational manipulation.'³⁷ The way that our environment affects our behaviours is known as the 'power of the situation'.

---

33 Phillip Zimbardo, *The Lucifer Effect: How Good People Turn Evil*. (Croydon: Rider Books, 2007), 274.
34 Peter Fromm, Douglas Pryer and Kevin Cutright. 'The Myths We Soldiers Tell Ourselves: and the Harm These Myths Do', *Military Review, September-October 2013* (2013), 60.
35 ADF, *Military Ethics*. (Australia: Directorate of Information, 2021), 34.
36 ICRC, *The Roots of Behaviour in War: Understanding and Preventing IHL Violations*. (Geneva: ICRC, 2004), 16.
37 Mastroianni, 'The Person-Situation Debate', 3.

The S-CALM Model

In 1971 Philip Zimbardo conducted an experiment simulating prison life at Stanford University, which was commissioned by the US Navy to investigate the effects of captivity on prisoners. Zimbardo aimed to show that situational rather than dispositional factors caused negative behaviour. Those taking part in the experiment were middle class students from across the US, who were randomly assigned to be guards or prisoners. A basement at Stanford University was converted into a prison with cells, solitary confinement room and a yard. The prisoners were given numbers, by which they were always referred to, prison uniforms and nylon stockings for their hair. The guards were given uniforms, reflective sunglasses, whistles, handcuffs, keys and clubs. They had complete control and were given no specific instructions, apart from maintaining order and not being allowed to use physical violence. The experiment was meant to last two weeks, but it was stopped after just six days. In a very short period the guards had become sadistic and the prisoners depressed and showing extreme stress. It showed that in the right situation people are susceptible to character changes. The experiment also demonstrated that the system of governance that allows the situation to develop is also a vital part in allowing evil acts to take place. Although Zimbardo's experiment has lost some of its credibility recently, it is still a useful pointer towards the power that a situation can have over individuals.[38]

Some of this power of the situation can be brought about by stress. Research has demonstrated that 'stress can affect our perception of a situation' and 'can cause tunnel vision, impaired hearing, and make events appear to be moving in slow motion.'[39] Research also suggests that the application of stressors can affect the way that personality functions as an individual applies a series of coping mechanisms. Stressors are simply a source of stress; stress itself can be defined as 'occurring when an individual is faced with demands that he or she finds impossible to satisfy.'[40] Those suffering stress are more likely to make cognitive errors, exhibit slow reaction times and have a deficit in judgement, whilst extended exposure can lead to a change in personality. Humans have evolved for a 'stress response to last about 30 seconds, enough to facilitate fight or flight. Evolution has not adapted

---

[38] For more information on the issues with Zimbardo's experiment see Richard Griggs 2014 paper.
[39] Deanna Messervey, Waylon Dean, Elizabeth Nelson, and Jennifer Peach, 'Making Moral Decisions Under Stress: A Revised Model for Defence', *Canadian Military Journal*, Vol. 21, No.2 (2021), 41.
[40] G. Breakwell and K. Spacie, *Strategic and Combat Studies Institute Paper 29: Pressures Facing Commanders*. (Camberley: Joint Services Command and Staff College, 1997), 4.

our brains or bodies to handle weeks or months of prolonged stress.'[41] The military are more likely to be exposed to stress than most professions. NATO notes that the military role creates stress which consists of three areas; 'role ambiguity (for example, uncertain of one's role in the workplace), role overload (for example, having many competing demands), and role conflict (for example, experiencing incompatible demands).'[42] The British Army recognises this and warns its leaders that 'the battlefield is anything but normal. The circumstances in which soldiers operate are among the most stressful found anywhere. Fatigue, hunger, revulsion and fear will impair judgement.'[43] Stress does not only affect combat troops; Breakwell and Spacie note that 'poor command relationships are potentially the single greatest source of stress. A commander needs to engender trust and confidence; if that is not present, the relationship with subordinates will become a source of friction and stress.'[44] For commanders also 'stress modulates risk taking, potentially exacerbating behavioural bias in subsequent decision making.'[45]

The three background concepts of the banality of evil, incremental steps and importantly the power of the situation all fall from the experiments and theories of Arendt, Milgram and Zimbardo. However, 'most military ethicists today use the findings of social psychology in both their research and in their training' and they form a good starting point from which my recent research has been conducted.[46]

Book Layout

Chapters 2 to 5 of this book outline how the power of the situation is enhanced and effects people's ability to act in an ethical way. In chapter 2 the situational influencers are explored. This chapter reveals my research into a series of case studies to identify the five common situational influencers, these are: hostile environment, normalised violence, weak leadership/lack of supervision, lack of resource/fatigue and enhanced emotional state. For each of these influencers an historical case study helps to highlight the effect

---

41 Stephen Flanagan, 'Losing Sleep', *Armed Forces Journal*, December 2011 (2011), 12.
42 NATO, *Leader Development for NATO Multinational Military Operations*, 116.
43 MOD, *Developing Leaders: A British Army Guide.* (Camberley, RMAS Academy Headquarters, 2014), 62.
44 Breakwell and Spacie, *Pressures Facing Commanders*, 20.
45 A. Porcelli and M. Delgado, 'Acute Stress Modulates Risk Taking in Financial Decision Making', *Psychological Science*, Vol. 20, No. 3 (2009), 278.
46 Peter Olsthoorn, 'Situations and Dispositions: How to Rescue the Military Virtues from Social Psychology', *Journal of Military Ethics*, August (2017), 91.

of it on those taking part. For hostile environment the murder of Baha Mousa, an Iraqi who died while in British Army custody in Basra, Iraq, in September 2003 will be used. For normalised violence, the Bloody Sunday massacre on 30 January 1972 in which British soldiers shot 26 unarmed civilians during a protest march in the Bogside area of Londonderry, Northern Ireland. The next influencer has been split, for weak leadership the Srebrenica massacre of 8,000 Bosniak Muslim men and boys in and around the town of Srebrenica, during the Bosnian War and for lack of supervision the beating of five Iraqis in May 2003 by British soldiers in Camp Breadbasket has been selected. Lack of resource has been applied to the rape and murder of a 14-year-old Iraqi girl, Abeer Qassim Hamza al-Janabi, and the simultaneous murder of her family by US Army soldiers. And finally, for enhanced emotional state, the Haditha Massacre which took place on 19 November 2005 in which US Marines killed 24 unarmed Iraqi civilians.

Chapters 3 and 4 cover the common behaviours that lead to unethical actions. The chapters will define and apply each of the 12 behaviours by using a case study. The cognitive behaviours have been divided into a cluster of six that mainly appear in individuals in chapter 3 and a further six that appear in a group setting in chapter 4.[47] The behaviours themselves are not unethical, but in the wrong situation can lead to unethical actions taking place. The six individual behaviours and case studies are: conformity / social comparison theory applied to the My Lai massacre in Vietnam in 1968, de-Individualisation applied to the British Hola Camp massacre in Kenya in 1959, obedience demonstrated by the British Amritsar massacre of 1919, cognitive dissonance applied to the British Batang Kali massacre of 1948, the bystander effect applied to the Canadian Somalia affair of 1993 and status quo bias as applied to the British trainee deaths at Deepcut Barracks between 1995 and 2002. The six group behaviours and case studies explored are: group think applied to the US Nisor massacre in Iraq in 2007, risky shift applied to the Nazi Police Battalion 101 in the Second World War, authority bias demonstrated by the British Chuka massacre of 1953, othering as demonstrated by the Serbian White Eagles during the Voćin massacre of 1991, dehumanisation applied to the US Army abuses in the Abu Ghrab Prison in 2003 and demonisation as applied to the US Army massacre at My Lai.

---

47 Note that there are in fact 13 theories, but social comparison theory and conformity have been combined for this book, as they are very similar concepts.

Ethical Leadership and Background Concepts

These chapters outlining the power of the situation are brought together in chapter 5 which describes the flow of unethical actions. It explains how, when a leader and their team find themselves in a stressful situation the power of the situation builds up and how they are susceptible to situational influencers. It continues to describe how these influencers in turn changes their behaviour and leads them to do unethical actions that they would not normally do if they were in a non-stressful situation and acting rationally. The chapter concludes by considering the British Sergeant Blackman case study to demonstrate the flow of unethical actions.

Having explored all this bad news, chapters 6 to 12 describe the S-CALM model and how these unethical actions can be avoided. When presenting the S-CALM model the military opinion as expressed in the doctrine of the British Army, US Army and Australian Defence Force has been included in the analysis of the model. Chapter 6 explains the way that the S-CALM model recognises and mitigates the five common Situational Influencers. It uses the case study of Major Richard Westley and B Company of the 1st Battalion the Royal Welch Fusiliers in Gorazde 1995. Chapter 7 illustrate how the S-CALM model recognises and mitigates the common individual behaviours. The six individual positive behaviour case studies are: social comparison theory applied to the crew of HMS *Coventry* in 1982, whilst for conformity the guardsmen who refused to take part in the Batang Kali massacre will be used. For deindividuation Sergeant Joe Darby's actions at Abu Ghraib will be used, for obedience the actions of General Mike Jackson during the Pristina Airport incident of 1999 will be explored. For cognitive dissonance Sergeant Franceschi's actions at Dien Bien Phu in 1954 will be explored, whilst for the bystander effect Warrant Officer Thompson actions at the My Lai massacre will be used and finally for status quo bias the activities of Police Battalion 101 will be investigated. Chapter 8 attends to how the S-CALM model recognises and mitigates the common group behaviours. The six group behaviour positive case studies are: for groupthink Trooper Griffin in Iraq in 2005, to demonstrate risky shift the conduct of some individuals in the Nisor massacre in Iraq will be used, for authority bias the decisions of Admiral Sandy Woodward during the Falklands War will be explored. For othering the actions of Lance Corporal Watson treating Afghans under fire in Afghanistan in 2010 will be investigated, for dehumanisation the Christmas Truce on the Western Front in 1914 will be used and finally for demonisation the actions of First Lieutenant Sibille in Belarus in 1941 will be used. Chapter 9 describes the role of Accountability in the S-CALM model and explores the different types of accountability and the role of the British Army Standards in maintaining culpability in stressful situations. It uses the case study of the Abu Gharib Prison abuse whistle blower, Sergeant Joe Darby, to demonstrate it. Chapter 10 presents the important part that leadership plays in the application of the S-CALM model and demonstrates this by exploring how the British Chief of the General Staff, General Sir Patrick Sanders, took the decision to remove the 3rd Battalion, The Parachute Regiment from deployment to Kosovo in 2022. Chapter 11 explores the role of virtue ethics and moral courage to guide a leader's moral compass. It uses the case study of how Warrant Officer Hugh Thompson stopped the My Lai massacre in 1968. The book concludes with Chapter 12 which pulls together all the research and presents how to apply the knowledge gained and use the STOP Protocol to avoid unethical actions in the future. It also explains in more detail the ethical leadership research conducted at RMAS in 2023, before presenting the Ethical Decision Making Coaching Tool which can be used on practical training exercises. It concludes by presenting the overarching ethical leadership hierarchy.

2

Situational Influencers

Introduction

In 2010, I was Commanding Officer of an infantry battalion. As part of some wider deployments, I had a platoon detached from my battalion to be part of a multinational security company in Kabul. They were responsible for the security of the multinational headquarters and for patrolling in the Kabul area. Although I spoke to the platoon commander often on the phone, it was exceptionally difficult to physically meet and offer advice and guidance. However, about half-way through the tour the Regimental Sergeant Major (RSM) and I found a week when we could get to Kabul and visit the platoon. The platoon commander and sergeant were both excellent soldiers and made a good team. They had initially found the friction of working in a multinational organisation difficult and we had discussed the management of these issues many times over a video link. However, when I visited it soon became clear that in an attempt to resolve this friction with the other nations, they had compromised some of their own standard operating procedures. These had not led to any negative issues to date, but there was an obvious slow erosion of the professional standards that we would have expected from them. The RSM and I as bystanders could look at their operations with fresh eyes and could see this shift in their performance. But it was difficult for the platoon staff to notice the effect the situation was having on them. I had an empathic chat with the platoon commander and explained that he was being influenced by the situation and had compromised some of the platoon's good practices and the RSM did the same with the platoon sergeant. They very quickly identified the issues we had noticed, such as the frustration of the operating environment, the weak leadership and the lack of personnel to perform all of their tasks and rapidly set about rectifying them. However more importantly, they established a new system for checking procedures between themselves that ensured that they were able to identify any lowering of the standards in the future and importantly prevent any ethical drift.

Chapter 1 demonstrated how early research established that the power of the situation influences the way people behave. Those early experiments also established 'just how insubstantial the situational factors that cause troubling moral failures seem to be.'[1] However in the military, leaders not only face routine stress but also 'the situational pressures of warfare [which] are, without doubt, very substantial indeed.'[2] Research has shown that there is nothing 'more dangerous than for commanders to convince themselves that they and their troops are somehow immune or invincible to situational factors.'[3] For a while, people have realised that there were these situational factors but most of the research in this area was on stress. However, it is important to establish what factors could enhance the power of the situation in a military environment other than stress. The Australian Defence Force identifies some of these factors when it states that 'when fear, fatigue, stress, injuries or grief are added, the burden of ethical decision-making is high.'[4] However there are other factors, that enhance the power of the situation and are known as situational influencers. In my case study research I wanted to ascertain what the common situational influencers were that build to influence ethical actions. My research analysed the reports, enquires and documents of over 20 historical case studies of poor ethical behaviour in the military. Each case study had influencers that enhanced the power of the situation, such as leadership, fatigue, emotions and the operating environment. This work identified that there were some common situational influencers that could be traced in most of the case studies. The research identified that there were common influencers that were key in these case studies that increased the pressure on leaders to behave in ways that they would not in normal situations. These were grouped together into five headings: hostile environment, normalised violence, weak leadership/lack of supervision, lack of resource/fatigue and enhanced emotional state, as displayed in Figure 2.1.

Although these five common situational influencers have been presented as separate items, they are in fact more complex than this and

---

[1] Dominic Murphy, 'From My Lai to Abu Ghraib: The Moral Psychology of Atrocity', *Midwest Studies in Philosophy* (2007), 34.
[2] Ibid.
[3] Paolo Tripodi, 'Understanding Atrocities: What Commanders Can Do to Prevent Them' in *Ethics, Law and Military Operations*, ed. David Whetham (London: Palgrave, 2010), 174.
[4] ADF. *Military Ethics*, 37.

Situational Influencers

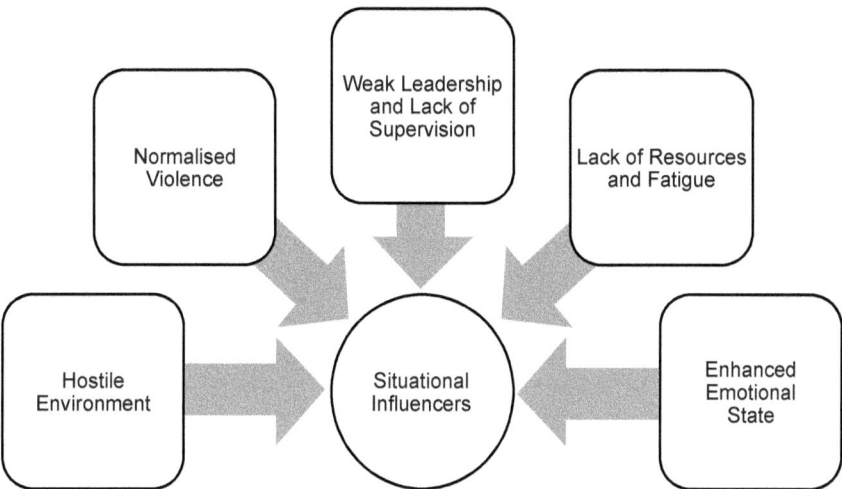

Figure 2.1 Situational Influencers

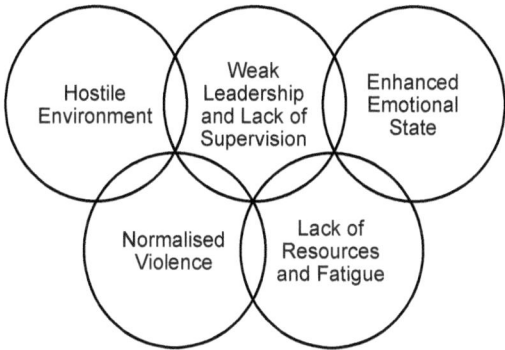

Figure 2.2 Overlap of Situational Influencers

there is a large degree of overlap between them. For example, a person who is fatigued is more likely to be in an enhanced emotional state. In a similar way, sustained exposure to a hostile environment can lead to normalised violence. An idea of how the common five situational influencers interlock and overlap is at Figure 2.2.

Each of the five situational influencers will now be explored and a supporting case study used to demonstrate how they can be seen to have taken effect in real life situations. In the majority of the case studies all five situational influencers were active, but the dominate one has been used to highlight its effect on those in the situation.

Hostile Environment

The first situational influencer to be explored is a hostile environment. As a simple definition, a hostile environment is one in which a person feels under threat or uncomfortable due to a perception of danger. A hostile environment therefore can be any situation from feeling uncomfortable on a city street on a Saturday night to the extreme of combat. People tend to behave differently in a hostile environment than in an atmosphere where they do not perceive a threat. A hostile environment produces stress which triggers the body's fight, flight or freeze response, a concept first described by Walter Bradford Cannon in the early 1900s. In turn this generates physiological changes 'such as blood pressure, heart rate and respiratory rate.'[5] These changes happen to allow for a person to run or fight at their maximum efficiency, commonly known as the fight or flight response. Further physical changes are made by the release of hormones which induce perspiration, tightening of the throat and a heightening of the ability to focus.

On operations, when soldiers are exposed to a hostile environment day after day, their harsh actions in a perceived hostile environment can soon become the acceptable social norm for the group. The need to be constantly alert and when required fight, rather than flight or freeze puts increased pressure on individuals. The longer an operational tour goes on the less aware soldiers become of their slip in behaviour in this enhanced state. Being immersed in a hostile environment for a long period of time can be closely linked to normalised violence. Very often when soldiers are eventually released from a hostile environment it can lead to an outpouring of emotional responses when given the opportunity.

Hostile Environment Case Study – Murder of Baha Mousa

Baha Mousa was an Iraqi who died while in British Army custody in Basra, Iraq, in September 2003. On 14 September 2003, Mousa, a 26-year-old hotel receptionist, was arrested along with six other men and taken to a British base. While in detention, Baha Mousa and the other captives were hooded, severely beaten and assaulted by a number of British troops. Two days later, Baha Mousa was found dead. A post-mortem examination found

---

5 Suresh Abhijit, Sai Swarna Latha, Pooja Nair and N. Radhika, 'Prediction Of Fight Or Flight Response Using Artificial Neural Networks', *American Journal of Applied Sciences*, Vol. 11, No. 6 (2014), 912.

Situational Influencers

that Baha Mousa suffered at least 93 injuries, including fractured ribs and a broken nose.

The inquiry into Baha Mousa's death found that it was caused by heat, exhaustion, fear, previous injuries and the treatment by the guards. Baha Mousa was hooded for almost 24 hours during the 36 hours he was held in custody by the 1st Battalion of the Queen's Lancashire Regiment. The Inquiry report noted that Baha Mousa was subject to several practices banned under both domestic law and the Geneva Conventions. The report also said a large number of soldiers assaulted Baha Mousa and that many others, including officers, must have known about the abuse. The report called Baha Mousa's death an episode of serious 'gratuitous violence.'[6]

Why did the soldiers of 1st Battalion the Queen's Lancashire Regiment abuse the detainees and beat Baha Mousa to death. In June 2003 they entered an extremely hostile environment? Troop levels 'fell drastically during the summer of 2003 from 26,000 to 9,000 to cover four provinces.'[7] The British policy had been for a soft approach and in this environment looting was rife in Basra and the security situation rapidly deteriorated. Just before the 1st Battalion deployed, six Royal Military Policemen were executed in their police station at Majar al-Kabir and just after they arrived, on 9 and 10 August, there were riots in Barsa as the looting had led to a breakdown in the provision of basic services such as electricity. In this turmoil the Shis majority mobilised behind a new leader, Moqtada al-Sadr. Shis militias were armed and financed by Iran and began to take control of areas and started to mount attacks on British troops.

The 1st Battalion the Queen's Lancashire Regiment started to take casualties from improvised explosive devices (IEDs). Then on 14 August 2003 a very popular officer, Captain David Jones was killed by an IED whilst he was traveling in an ambulance. Around the same time, three Royal Military Police soldiers were killed whilst on patrol. Research has demonstrated that 'The British seemed unable to stop the mounting violence and increasingly became the focus of attacks by the militias.'[8] These acts confirmed that Basra had become a hostile environment and this hardened the attitudes of the troops toward the locals and led to a crackdown on suspected insurgents in the city. A month after Captain Jones's murder, on Sunday 14 September 2003,

---

6 William Gage, *The Report of the Baha Mousa Inquiry*. (London: The Stationery Office, 2011), 380.
7 Warren Chin, 'Why Did It All Go Wrong? Reassessing British Counterinsurgency in Iraq', *Strategic Studies Quarterly Winter 2008* (2008), 128.
8 Chin, 'Why Did It All Go Wrong? Reassessing British Counterinsurgency in Iraq', 132.

the 1st Battalion launched Operation Salerno, which saw soldiers searching a series of hotels where they believed insurgents were working from. At the Ibn Al Haitham Hotel they discovered weapons, military clothing and fake ID cards, and they arrested ten men, one of whom was the hotel receptionist, Baha Mousa. Troops in the 1st Battalion believed that those arrested had been involved in the deaths of Captain Jones and the Military Police soldiers. Three months of being exposed to the hostile environment made them want revenge and they behaved in ways that they would not normally have done.

Normalised Violence

The second common situational influencer can be linked to a hostile environment and is termed normalised violence. A definition of normalised violence is: 'the acceptance that violence is an immutable part of life, that depictions of violence do not have real life consequences, and that it is the responsibility of the victim, not the perpetrator, to prevent violence.'[9] In a military situation, it has been noted that 'violence becomes a normal tool to use even in situations it would not be during peace time.'[10] Everyone has exposure to some level of violence, but for most people their violence threshold is low as they are rarely exposed to violence and even small acts of violence can cause a strong reaction. However, if people are exposed to regular violence it can be seen as an immutable part of our daily life, it is then soon accepted as the social normal. In this situation the threshold of violence is heightened and it takes more and more violence to cross this new threshold and for it to register as an issue.

In intense, combat heavy deployments, soldiers can quickly become immune to the violence around them. Death and injury become the norm and therefore their threshold to inflicting pain on others is lowered. In Albert Bandura's work on moral disengagement, he wrote that 'violence is made morally acceptable by claiming that one's injurious actions will prevent more human suffering than they cause.'[11] This justification of violence makes it more acceptable to both the individual and the group. The longer that violence is a part of daily life, such as on a six month operational deployment

---

9 Social Norms toolkit: The Normalization of Violence, WCASA. https://www.wcasa.org/wp-content/uploads/2020/01/PDFforToolkitNormalizationofViolence.pdf.
10 Armouring Against Atrocity, King's College London, Centre for Military Ethics, 2016. https://militaryethics.uk/en/course/library.
11 Albert Bandura, 'Moral Disengagement' in *The Encyclopaedia of Peace Psychology, First Edition*, ed. Daniel Christie (New Jersey: Blackwell Publishing, 2012), 2.

where casualties are taken, the more likely that soldiers will see violence as the norm. In this state the 'range of acceptable behaviour becomes so wide and there is no clear moral reference point for right and wrong.'[12] This can lead to what is known as 'opportunity without consequence: the idea that wartime is not governed by the same societal norms or restraints.'[13] When soldiers believe this idea they commit acts that in normal times would be unacceptable but which they now recognise as the social norm.

Normalised Violence Case Study – Bloody Sunday Killings

Bloody Sunday was a massacre on 30 January 1972 when British soldiers shot 26 civilians during a protest march in the Bogside area of Londonderry, Northern Ireland. 14 people died: 13 were killed outright, while the death of another one four months later was attributed to injuries on the day. The march had been organised by the Northern Ireland Civil Rights Association (NICRA) to protest against internment without trial. The soldiers were from the 1st Battalion the Parachute Regiment.

The 1st Battalion the Parachute Regiment were not based in Londonderry, but in Belfast. They had first deployed to Northern Ireland for a four month tour of duty between October 1969 and February 1970. They then returned to Belfast on a two year residential tour in September 1970. This was a very violent time in Northern Ireland with regular riots, shootings and bombing attacks. The 1st Battalion were involved in Operation Demetrius on 9/10 August 1971 which was the interment of individuals believed to be involved in insurgent activities. Around 342 people were arrested in the first wave of internment. This action created a great deal of violence on the streets and around 7,000 people either fled or were forced out of their homes in heightened sectarian attacks. During the whole of Operation Demetrius, 20 civilians, two IRA members and two British soldiers were killed. 11 of the civilians were killed by the 1st Battalion the Parachute Regiment in what became known as The Ballymurphy Massacre. Therefore, before the battalion was even deployed to Londonderry they had been exposed to excessive bloodshed.

The area known as 'Free Derry' was a 'no go' area for the British Army units in Londonderry and the 1st Battalion the Parachute Regiment was redeployed from Belfast to Londonderry to show that there were no areas that they could not operate in and make arrests during the riots that the

---

12 King's College London, Centre for Military Ethics, *Armouring Against Atrocity*.
13 Ibid.

authorities knew would happen after the civil rights march. These riots broke out in the late afternoon when the march was breaking up. At this stage there was some shooting between some Official IRA gunmen and some Parachute Regiment soldiers and other soldiers claimed to have been fired at, but with no later evidence. Baton rounds and CS Gas were also fired by the units holding the barriers which surrounded the rioters. Ten minutes or so after the shooting incident, Support Company were ordered in to make some arrests. The Mortar Platoon of Support Company deployed deep into the 'no go' area and shortly after their leader, Lieutenant N, fired some shots in the air. He said that this was to prevent some civilians from 'attempting to rescue a man who had been arrested by one of the soldiers' he went on that they 'were throwing stones and similar missiles at the soldiers', he concluded that he did this to prevent 'the crowd from attacking him and the soldiers with him.'[14] Once these shots were fired, the real shooting began, with 21 soldiers firing 108 rounds. At the end of this, 13 civilians were dead. Before these killings, the soldiers had been exposed to over 16 months of violence, which had slowly reduced their threshold of violence and allowed them to act in a way that they most likely considered was an 'opportunity without consequence'. Once shots were fired, there was a contagion and soldiers acted in a way that in normal times would be unacceptable but which they now identified as the social norm.

Weak Leadership and Lack of Supervision

The third influencer is weak leadership and lack of supervision. The British Army defines leadership as 'a combination of character, knowledge and action that inspires others to succeed.'[15] Leadership is vital in all military situations and a leader has a profound effect on the behaviours of their followers. The case study research has demonstrated that weak leadership is normally seen when a leader uses a hands-off style with their team when in fact their team needs to be given more guidance and direction. This style, known as laissez-faire leadership is defined as a style in which a leader 'abdicates responsibility, delays decisions, gives no feedback, and makes little effort to help followers satisfy their needs.'[16] In this situation

---

14 Lord Saville, *Principal Conclusions and Overall Assessment of the Bloody Sunday Inquiry*. (London: The Stationery Office, 2010), 22.
15 MOD, *Army Leadership Doctrine*. (London: HMSO, 2021), 1-2.
16 Peter Northouse, *Leadership, 9th ed.* (London: Sage, 2022), 196.

the person in charge has become a non-leader who avoids making decisions and taking responsibilities, refuses to take a stance and shows a lack of interest in the outcome. However, weak leadership can also be the opposite in that a leader micro-manages their team and followers feel they have no responsibility for their actions. In this case followers feel that they can act in an unethical way, as the leader is responsible for their actions. Weak leaders often do not display the correct level of supervision. Supervision is defined as 'the act of watching a person or activity and making certain that everything is done correctly.'[17] This lack of supervision can allow soldiers to become swayed by some of the other situational influencers and this again can lead to ethical drift. Ethical drift has been defined as 'an incremental deviation from ethical practice that goes unnoticed by individuals who justify the deviations as acceptable and who believe themselves to be maintaining their ethical boundaries.'[18] The Royal Navy Submarine Service term ethical drift as ethical corrosion. They believe that 'Ethical Corrosion is the deterioration of: 1. Our desire to be ethical. 2. Our ability to discern accurately what is ethical. 3. Our willingness to challenge what is unethical – generating a 'walk on by' culture.'[19] Ethical drift escalates until even serious breaches of the ethical code are rationalised. A simple example of this is not following Standing Operating Procedures or wearing the incorrect dress. It has been shown that 'studies demonstrate that a nominal clothing manipulation can have effects on the behaviour of the wearer.'[20] Therefore a leader needs to be able to control the ethical direction that their commands take and intervene in ethical drift before it spirals out of control.

Weak Leadership Case Study – Srebrenica Massacre

The Srebrenica massacre occurred in July 1995 when more than 8,000 Bosniak Muslim men and boys were killed in the town of Srebrenica during the Bosnian War. Prior to the massacre, United Nations (UN) had declared the besieged enclave of Srebrenica a safe area under UN protection. The

---

17 Cambridge Dictionary. https://dictionary.cambridge.org/example/english/lack-of-supervision.
18 Carole Kleinman, 'Ethical Drift When Good People Do Bad Things', *JONA's Healthcare Law, Ethics, and Regulation*, Vol. 8 No. 36 (2006), 72.
19 MOD. A Practical guide to Military Ethics within the Submarine Service. (London: HMSO, 2019), 18.
20 Johnson et al, 'Dress, Body and Self: Research in the Social Psychology of Dress', *Fashion and Textiles 2014*, Vol. 1, No. 20 (2014), 8.

area was protected by a force of 370 lightly armed Dutch soldiers who were commanded by Lieutenant Colonel Thom Karremans.

The Serb offensive against Srebrenica began on 6 July 1995. Some of the Dutch soldiers retreated into the enclave after their posts were attacked and some surrendered to the Serbs as they believed that 'Karremans had not displayed adequate leadership.'[21] In the evening, Lieutenant Colonel Karremans was filmed drinking a toast with General Mladić, the Serb commander, during an attempt to negotiate. A week later Dutch troops witnessed Serb soldiers murdering some of the Bosniaks but took no action. On Friday, 21 July 1995, Lieutenant Colonel Karremans agreed that he would withdraw the Dutch soldiers from Srebrenica rather than continue to protect the Bosniak occupants. When he departed, Karremans accepted gifts from General Mladić, smiled and shook hands. The Serbs then moved in and started the massacre of the Bosniak men and boys. In his review of the incident, P.J. de Vin concluded that 'Karremans did not take the correct moral and ethical approach as he decided not to defend the enclave during the attack on Srebrenica.'[22] During the whole operation, Karremans showed weak leadership of the battalion.

Lack of Supervision Case Study - Camp Breadbasket Abuse

In May 2003, British soldiers of the Royal Regiment of Fusiliers took part in an operation nicknamed Operation Ali Baba which was aimed at stopping looting from their camp. In the early hours of the 15 May, British soldiers rounded up local Iraqis from in and around Camp Breadbasket. Five Iraqis were unlawfully detained, assaulted and beaten using car aerials and wooden sticks and a couple of victims were placed on a forklift truck. Some of them were made to strip naked and forced to participate in sexually humiliating/ abusive acts, which were photographed by many soldiers. Three soldiers were court martialled: Corporal Kenyon, Lance Corporal Cooley and Lance Corporal Larkin for the abuse.

The Royal Fusiliers were at a loss for what to do with the looters once they had rounded them up. The Battalion Quartermaster, Major Dan Taylor, was in charge of Camp Breadbasket where the detainees were taken, and he gave the ambiguous order to 'work them hard'. At the court martial 'Maj

---

21 P.J. de Vin, 'Srebrenica, the impossible choices of a commander', Master Thesis, United States Marine Corps Command and Staff College (2008), 10.
22 de Vin, 'Srebrenica, the impossible choices of a commander', iv.

Taylor admitted giving the order to work detainees hard', the prosecution lawyer contended that 'it is precisely your order that resulted in these three soldiers being charged on very serious offences.'[23] Having given this order, Major Taylor neither checked that the order was being carried out as he intended, nor did any other person in authority. One of the accused, Corporal Kenyon, said at the court martial that 'there had been a breakdown in the chain of command at Camp Breadbasket, adding that the way the camp was run was "infected".'[24] He did not explain what he meant by infected, but it was obvious that there was a lack of supervision, which allowed for ethical drift by junior non-commissioned officers who were allowed to enter the spiral violence which led from the initial supervision of litter picking duties to sexual and physical abuse.

Lack of Resource and Fatigue

The case study research revealed how in many situations, soldiers believe that they do not have the correct level of resource to complete the task set them. This maybe in the amount of materiel, such as food, water, equipment or weaponry. However, in many case studies the two key resources lacking were time and people. Leaders found that they did not have enough time to do things as they should and had to prioritise what actions they could achieve in the limited time given. This could lead to corners being cut and activities not being conducted properly. There were also issue with soldiers not being trained properly due to a lack of time to do so. In the 2002 a study by the US Army War College Strategic Studies found that out of a 256 training days available a year, it would have taken company commanders a total of 297 days, a deficit of 41 days, to accomplish all assigned training.'[25] Recent NATO doctrine confirmed that 'time pressure has also been shown to negatively affect moral decision making, with those put under such conditions being less likely to exhibit ethical behaviour.'[26] It was similar in personnel; leaders found that they did not have enough people to conduct actions as they would have wanted, so people were stretched and corners cut to meet outputs.

---

23 Officer's orders 'led to abuse', BBC News, 21 January 2005, http://news.bbc.co.uk/1/hi/uk/4193751.stm.
24 Trial highlights camp's problems, BBC News, 23 February 2005. http://news.bbc.co.uk/1/hi/uk/4287449.stm.
25 No Time, Literally, For All Requirements, Crispin Burke. *https://www.ausa.org/articles/no-time-literally-all-requirements.*
26 NATO, *Leader Development for NATO Multinational Military Operations*, 9-14.

When there is a deficiency of people and time it invariably leads to fatigue amongst those in teams. Fatigue is defined as 'a state of tiredness and diminished functioning. Fatigue is typically a normal, transient response to exertion, stress, boredom, or inadequate sleep.'[27] Sleep is the 'rest and recovery from the wear and tear of wakefulness' and is an essential part of our existence.[28] It is universally accepted that the human body requires eight hours sleep to regenerate and that prolonged periods of less than this reduces cogitative ability. In fact, junior officers who are normally in their mid-20s, 'require from 8.50 to 9.25 hours sleep per night.'[29] Sleep deprivation can have a wide range of effects on these functions including, poor attention, deficient memory, mood swings and inferior decision making. Studies into sleep patterns during combat operations suggest that the amount of sleep gained per night is less than that recommended. As part of a ten year study into the role of sleep in the military, the US Army was examined during the invasion of Iraqi in 2003. In this survey up to 'eighty three percent of respondents showed moderate sleep deprivation, whilst up to twenty three percent showed significant symptoms, with an average of six point six seven hours sleep, reducing to four hours during combat.'[30] Even with this level, '34% reported falling asleep at least once when they should have been awake.'[31] The Norwegian Army conducted a sleep deprivation trial over five nights with young officers. The aim of the experiment was to test if stress and lack of sleep would result in subjects firing live ammunition at targets which were real people rather than dummies. In the experiment:

> 59% of the students (1st or 2nd lieutenants) fired their weapons and 41% did not. All of the 41% who did not fire said they noticed people in the target area and therefore did not fire, but only one of them tried to warn the others to stop firing.[32]

The results demonstrated how the majority opened fire and even those that did not did not warn the others. The Norwegian Army concluded that

---

27 APA Dictionary of Psychology. https://dictionary.apa.org.
28 Nita Miller, Panagoiotis Matsangas and Aileen. Kenney, 'The role of Sleep in the Military: Implications for Training and Operational effectiveness', in *The Oxford Handbook of Military Psychology*, eds. J. Laurence and M. Matthews (New York: Oxford University Press, 2012), 263.
29 Miller, Matsangas and Kenney, 'The role of Sleep in the Military', 264.
30 Breakwell and Spacie, *Pressures Facing Commanders*.
31 Miller, Matsangas and Kenney, 'The role of Sleep in the Military', 277.
32 Rolf Larsen, 'Decision Making by Military Students Under Severe Stress', *Military Psychology*, Vol. 13, No.2 (2001), 89.

'although simple and well-learned physical tasks such as weapons handling suffer little, there seems to be a rapid general degeneration in the cognitive or psychological area.'[33] In another study conducted during a 54 hour military exercise, 'command performance dropped by thirty percent after the first night without sleep and by sixty percent after the second.'[34] Other experiments wanted to find out how quickly it would take for soldiers cognitive ability to degenerate. They found that 'after four hours sleep for six nights, participants' performance was just as bad as those who had not slept for twenty-four hours straight.'[35] The study then identified that 'after ten days of just seven hours of sleep, the brain is as dysfunctional as it would be after going without sleep for twenty-four hours.'[36] Despite this, the military are reluctant to believe such findings, asserting that motivation and determination can overcome fatigue and that 'senior officers imagined that they did not need eight hours sleep.'[37] Nevertheless, the fact remains that evolution has not prepared humans for prolonged periods of stress brought on by a lack of sleep. When a leader is in this fatigued state there is a 'strong impairment of their ability to activate autonomous and principle-oriented moral reasoning.'[38] Whilst it has also been demonstrated that 'insufficient sleep has also been linked to aggression, bullying, and behavioural problems.'[39] Finally, a 'lack of sleep may promote unethical behavior [sic] by diminishing self-control'.[40] Therefore, when soldiers feel under resourced it can lead to a degree of hopelessness. This feeling can in turn lead soldiers to not care about what they do. Leaders who lack sleep can find it difficult to control this behaviour and make poor cognitive decisions.

Lack of Resource and Fatigue Case Study – Yusufiyah Killings

On 12 March 2006, five Soldiers from 1st Platoon, B Company, 1st Battalion, 502nd Infantry Regiment of the 101st Airborne Division abandoned their

---

33 Ibid., 90.
34 S. Flanagan, 'Losing Sleep', *Armed Forces Journal, December 2011.* (2011).
35 Matthew Walker, *Why We Sleep.* (London: Penguin, 2017), 136.
36 Ibid., 140.
37 Miller, Matsangas and Kenney, 'The role of Sleep in the Military', 280.
38 O. Olsen, S. Pallesen and J. Eid, 'The Impact of Partial Sleep Deprivation on Moral Reasoning in Military Officers', *SLEEP 2010,* Vol. 33, No. 8 (2010), 1089.
39 Matthew, *Why We Sleep,* 148.
40 Christopher Barnes, John Schaubroeck, Megan Huth and Sonia Ghumman, 'Lack Of Sleep And Unethical Conduct', *Organizational Behavior and Human Decision Processes,* Vol. 115 (2011), 177.

posts and headed to the village of Yusufiyah. The soldiers, Sergeant Paul Cortez, Specialist James Barker, Private Jesse Spielman, Private Steven Green and Private Brian Howard, who served as the group lookout, committed the gang-rape and murder of a 14 year old Iraqi girl, Abeer Qassim Hamza al-Janabi. They also murdered her mother, Fakhriah, her father, Qassim and her six-year-old sister Hadeel.

The overstretch of B Company began even before they deployed to Iraq. The US Army knew that they did not have enough troops in the country to quell the increasing violence and in 2007 George W. Bush got a bill through the US Government which became known as the Surge and greatly increased US combat power. Therefore, when the 1st Battalion, 502nd Infantry Regiment deployed they were detached from the 101st Airbourne Division and sent to the 3rd Division and to operate in the south of Bagdad where more troops were needed. Colonel Todd Ebel, the brigade commander felt that he was under resourced and discussed the new deployment with the 3rd Division Deputy Commander, Major General Webster who told Ebel that 'no one really cares about South Baghdad.'[41] Colonel Ebel told them that he believed that he was 'about a US brigade short, if not less than an Iraqi brigade or two short' in the post tour report he confirmed that 'history has proven that because there are now two brigades under the 3d ID [3rd Infantry Division] in MND-Central [Multi National Division-Central].'[42]

Into this environment, Lieutenant Colonel Kunk, the battalion commanding officer, deployed B Company to six static Traffic Control Points (TCP) on a road known as Route Sportstar. They also deployed a platoon to set up a patrol base at the bridge over the River Euphrates in an old water treatment plant, the troops referred to the Jurf al Sukr Bridge as the Alamo Bridge. B Company was now over stretched. It has been assessed that 'B Company's responsibility was too great for its strength. Fire teams were doing squads' jobs, squads doing platoons' jobs, and platoons doing companies' jobs. A TCP of four soldiers, 24 hours a day, did not have enough troops to do a proper guard rotation.'[43] The troops were getting very fatigued and turned to alcohol and drugs to cope. It has been assessed that 'by mid

---

41 John Cushman, Chain of Command Performance of Duty, 2d Brigade Combat Team, 101st Airborne Division, 2005-06. *A Case Study Offered to the Center for Army Professional Ethic* (2011), 10.
42 Ibid., 11.
43 Ibid., 17.

December it should have become clear to Lieutenant Colonel Kunk that B Company was stretched too thin and that a correction must be made.'[44]

In the first three months, B Company lost three platoon leaders, a first sergeant, a squad leader and a team leader to death or injury. In December Captain Goodwin, B Company's commander, broke down and cried when Lieutenant Ben Britt the 1st platoon commander was killed in an IED attack. In February, a new platoon sergeant, Sergeant First Class Fenalson was appointed. Although he was a strict disciplinarian, he rarely left the base to visit the TCPs which allowed for ethical drift to happen and as earlier described the use of alcohol and drugs to become common place. The lack of people and fatigue ultimately became situational influencers that caused members of B Company to behave in ways they would not normally have considered.

Enhanced Emotion State

In situations where soldiers face intense emotions, their ability to do the right thing is often reduced. In combat situations judgement can be clouded by heightened emotions as the 'extreme situation of war can lead to extreme emotions e.g. rage and fear acting as a barrier to normal cognitive decision making.'[45] It has been shown that 'emotional reaction at the moment [of stress] is more influential in determining choice than the rational evaluation of options that may have been conducted beforehand.'[46] Another researcher has suggested that 'reason behaves like a lawyer who devises arguments after the fact to justify positions that have been taken based on intuition and emotions.'[47] Therefore when under a stressful situation emotion and not reason can affect the way soldiers behave. These emotions or visceral states have been shown to be 'strong physiological influences' and consist of 'anger, disgust, sexual arousal, thirst, and hunger.[48]

The list of emotions induced in a military context has been researched by many people and include 'intense feelings of euphoria, regret, grief, anger,

---

[44] Ibid., 23.
[45] King's College London, Centre for Military Ethics, Armouring Against Atrocity.
[46] Olaoluwa Olusanya, *Emotions, Decision-making and Mass Atrocities*. (Farnham: Ashgate, 2014), 116.
[47] Edgar, Karssing, 'The E-Word (Emotions) in Military Ethics Education: Making Use of the Dual-Process Model of Moral Psychology' in *Violence in Extreme Conditions: Ethical Challenges in Military Practice*, ed. Eric-Hans Kramer and Tine Molendijk (Cham: Springer, 2023), 132.
[48] Deanna Messervey, Waylon Dean, Elizabeth Nelson, and Jennifer Peach, 'Making Moral Decisions Under Stress: A Revised Model for Defence', *Canadian Military Journal*, Vol. 21, No. 2 (2021), 41.

or disgust.'[49] Others have noted 'hunger, extreme temperature, time pressure, pain, fear of dying, and the fear of seeing fellow personnel killed in action.'[50] Whilst a US study of soldiers in Iraq in 2006 noted that 'anger was associated with unethical warzone behaviour.'[51] Canadian research has shown that 'stressors associated with peace support missions can generate a simmering undercurrent of emotions, from frustration and anger, to despondency, guilt and revenge.'[52] Rage is often felt when there is a feeling that soldiers want to take revenge for a previous situation. Especially in counter insurgency operations there is a feeling of comparative morality, in that 'they do it' so why shouldn't we? This links to frustration, with soldiers wanting fair combat with the enemy, rather than taking casualties without a chance to engage the enemy. There is research to show that in these situations there is 'the potential for frustration to be translated into violence.'[53]

Under the stress of the situation these stressors cause behaviours that they would not want to do if they were thinking in a rational manner. The Australian Defence Force noted that 'when fear, fatigue, stress, injuries or grief are added, the burden of ethical decision-making is high.'[54] Research has also demonstrated that when emotionally charged these 'visceral states can increase the risk of unethical behaviour because they can lead to heat-of-the-moment thinking, where we are tempted to satisfy our immediate desires at the expense of our long-term goals.'[55] It has also been shown that 'people miss or misperceive a lot of things at the extremes of arousal.'[56] This short termism is typical of intuitive, System 1, thinking with people being driven by their emotions and immediate needs to act without stopping to consider what the consequences might be, whether that action is frustration, anger or fear. This in turn leads to unethical behaviour, research has shown that 'emotions also shape behaviour, including the behaviour of actors in a war.'[57]

---

[49] Human Behavior in Military Contexts, National Research Council, Washington, DC: The National Academies Press, 2008, https://doi.org/10.17226/12023, 17.
[50] Messervey, Dean, Nelson, and Peach, 'Making Moral Decisions Under Stress', 41.
[51] Magnus Linden et al, 'A latent core of dark traits explains individual differences in peacekeepers' unethical attitudes and conduct', *Military Psychology*, Vol. 31, No. 6 (2019), 500.
[52] George Shorey, 'Bystander Non-Intervention and the Somalia Incident', *Canadian Military Journal, Winter 2000-2001* (2001), 27.
[53] Mastroianni, 'The Person-Situation Debate', 7.
[54] ADF, *Military Ethics*, 35.
[55] Messervey, Dean, Nelson, and Peach, 'Making Moral Decisions Under Stress', 41.
[56] Leo Murray. *Brains & Bullets: How Psychology Wins Wars*. London: Biteback, 2013, 193.
[57] Samuel Zilincik, 'The Role of Emotions in Military Strategy', *Texas National Security Review*, Vol. 5, Issue 2 (2022), 15.

Enhanced Emotions State Case Study – Haditha Massacre

The Haditha massacre took place on 19 November 2005, in which US Marines killed 24 unarmed Iraqi civilians. The killings occurred in the city of Haditha. Five men were taken from a taxi and killed; the remainder were killed in three houses that were attacked. Among the dead were five women and four children still wearing their pyjamas and in bed, they were shot multiple times at close range. The 3rd Battalion, 1st Marines were on their third tour of Iraq since the invasion in 2003, when they deployed in September 2005. On the morning of the massacre, K Company deployed a routine vehicle patrol under Staff Sergeant Frank Wuterich. They knew this was a routine, but dangerous patrol as the week before three Marines had been injured by an IED on the route they were due to take. As they passed the area of the previous attack, they were attacked with an IED made of buried artillery shells. This killed Lance Corporal Miguel Terrazas and injured two other marines, Lance Corporal James Crossan and Private Salvador Guzman. James Crossan told ABC News, 'we are pretty much like one family, and when your teammates do get injured and killed, you are going to get pissed off and just rage.'[58] The death of Miguel Terrazas created emotions of both anger and revenge in the troops. Therefore, when a taxi arrived on the scene just after the event, it was stopped for searching as was normal procedure. However, the four teenagers and the taxi driver were all shot dead by Sergeant Wuterich. After their deaths, the platoon commander, Lieutenant William Kallop, arrived on the scene. He was also emotional about the death of Terrazas and believed he was taking small-arms fire. Kallop then gave the ambiguous order "to take the house" that he believed the fire was coming from. Nineteen of the civilians killed were in the three adjacent houses which US Marines entered, employing grenades and small arms. No insurgents or weapons were found in the two houses and an old AK 47 was found in the third.

One of the soldiers that took part in the massacre, Sergeant Sanick Dela Cruz, believed that it 'was a revenge driven massacre.'[59] Later on the day of the massacre he urinated on the head of one of the teenagers who were killed from the taxi. He said that it was an 'expression to the anger he had held inside him all day' he later explained he did it as 'you're mad; you're angry

---

58 Micheal Duffy, Tim McGirk, and Bobby Ghosh, 'The Ghosts of Haditha', *TIME*, Vol. 167, No. 24 (2006), 28.
59 'Witness at Haditha', Bryan Smith, *Chicago Magazine, July 2008* (2008). https://www.chicagomag.com/Chicago-Magazine/July-2008/Witness-at-Haditha/.

over what had happened.'⁶⁰ Canadian Army research has shown that 'anger, a powerful visceral state, was salient in several high-profile cases of unethical behaviour by military personnel' it added that 'U.S. Marines killed unarmed non-combatants in Haditha after an improvised explosive device killed one of their members.'⁶¹ At Haditha a well trained and experienced group of soldiers allowed their emotions, especially anger, to drive their actions and behave in ways they were all later to regret.

Summary of Situational Influencers

Although the case study research demonstrated there were five clear common situational influencers in the case studies, there are of course other influencers, and each case study was slightly different. The situational influencers do not operate in isolation, there is a lot of overlap and interaction between them and this can be quite complex. For example, a hostile environment can lead to normalised violence and when people are fatigued, they are more likely to be emotional. The situational influencers are present in all stressful situations and they work to enhance the power of the situation and cause people to behave in ways they would not have done in a non-stressful situation and thinking in a rational way. Chapter 3 will now investigate six common individual behaviours.

---

60 Bryan Smith, 'Witness at Haditha.
61 Messervey, Dean, Nelson, and Peach, 'Making Moral Decisions Under Stress', 41.

3
Common Individual Behaviours

Introduction

When I first arrived in the battalion from the Royal Military Academy Sandhurst in the early 1980s, I experienced two initiation ceremonies. One in the Officer Mess which involved a lot of alcohol and one in my platoon which involved a lot of genitalia. I accepted this maltreatment as a tradition that all new people joining the organisation underwent and although slightly put out by them, I did not question them. However, in time the status quo was questioned, why did these things happen to new people? They were soon seen not as traditions but as abuse of young people and they were stopped throughout the British Army.

There are around 182 different cognitive behaviours. In this book, cognitive behaviours refer to how individuals and groups think and respond to the situation that they find themselves. Most of these cognitive behaviours are related to fast thinking which can result in the use of heuristics, which in turn can lead to biases, all of these terms need to be explained before investigating how they are implemented in stressful situations. In 1921 Carl Jung the Swiss psychiatrist stated that people tend to be either extrovert or introvert thinkers. The extrovert thinker tends to be intuitive, whilst the introvert tends to be more rational. In the 2011 book, *Thinking, Fast and Slow*, Daniel Kahneman updated these ideas and called intuitive thinking System1 thinking. He stated that 'System 1 operates automatically and quickly, with little or no effort and no sense of voluntary control.'[1] He concluded that

---
1 Daniel Kahneman, *Thinking, Fast and Slow*. (New York: Farrar, Straus and Giroux, 2011), 22.

'System 1 operates automatically and cannot be turned off at will, errors of intuitive thought are often difficult to prevent.'[2] These errors in intuitive thought are brought about by System 1's use of cognitive shortcuts known as heuristics. 'Heuristics are cognitive 'rules of thumb' that allow us to make quick mental calculations that are necessary for quick decisions and responses.'[3] Most people use heuristics and they are a normal part of our cognitive process and work successfully for most of the time giving adequate answers to daily problems. However, they can sometimes lead to biases, which can be 'described as systematic, predictable errors of judgement.'[4] These biases are not the conscious bias where one might be prejudiced against an individual or group such as sexism or racism but are unconscious biases. With unconscious bias a person is more likely to 'give more weight to information that leads us in a desired direction'.[5] This mix of System 1 thinking, heuristics and biases all form the common cognitive behaviours referred to in this book.

The UK, US and Australian armed forces have identified in their doctrine that cognitive behaviours need to be considered when taking decisions. The British Army states that 'leaders who set a bad example are dangerous, as inappropriate behaviour can be infectious and can quickly become the norm if not corrected.'[6] This issue of contagion is important and will be identified later in the text on conformity, social comparison theory and groupthink. The UK also note the fallout of unethical behaviours, stating that 'unacceptable behaviour undermines trust and cohesion, directly impacting operational effectiveness.'[7]

The US Army agrees that 'all leaders are susceptible to displaying counterproductive leadership behaviors [sic] in times of stress, high operational tempo, or other chaotic conditions.'[8] There is here the understanding that the power of the situation can lead to the display of common behaviours. Finally, the Australian Defence Force affirm that 'individuals will be held personally accountable for their own wrongdoing

---

2 Ibid., 30.
3 MOD, *Joint Doctrine Publication 04 Understanding and Decision-making*, 2nd ed. (London: DCDC, 2016), 40.
4 Daniel Kahneman, Oliver Sibony and Cass Sunstein, *Noise: A Flaw in Human Judgement*. (London: William Collins Books, 2022), 161.
5 MOD, *Joint Doctrine Publication 04 Understanding and Decision-making*, 65.
6 MOD, *Army Leadership*, 2-9.
7 MOD, *The Army Leadership Code: An Introductory Guide*. (Camberley: The Centre for Army Leadership, 2015), 10.
8 US Army, *Army Leadership and The Profession*, 8-7.

in war, as in peace, regardless of the behaviour of the people around them.'⁹ Accountability will be explored later, but the Australians here are referring again to concepts of conformity, social comparison theory and groupthink. However, there is also a link to behaviours such as deindividuation and risky shift.

Importantly, it should be noted that these behaviours are not in themselves unethical, but under the power of the situation, when things go wrong they can lead to unethical actions taking place. As with the situational influencers, these behaviours are complex and interactive, however for ease, each will be investigated by using a case study to get a deeper understanding of how that behaviour appears on military operations. Using real life situations for these investigations of 'the experience of soldiers that have participated in atrocities' allows for a better understanding of 'their behaviour.'¹⁰ The cognitive behaviours have been divided into six common individual and six common group behaviours. The first group of six common cognitive behaviours are those that mainly appear in oneself and are covered in this chapter. The six individual behaviours and case studies are: conformity/ social comparison theory applied to the My Lai massacre in Vietnam in 1968, de-Individualisation demonstrated by the British Amritsar massacre of 1919, obedience applied to the British Hola Camp massacre in Kenya in 1959, cognitive dissonance theory applied to the British Batang Kali massacre of 1948, the bystander effect applied to the Canadian Somalia affair of 1993 and status quo bias applied to the British trainee deaths at Deepcut Barracks between 1995 and 2002.¹¹

Social Comparison Theory and Conformity

Social comparison theory and conformity are two separate theories but they have been grouped together here as they are extremely similar. They tend to be triggered by similar situations and produce similar unethical actions. Social comparison theory was first discussed by Leon Festinger, an American social psychologist in 1954. He found that 'self evaluation and the necessity for such

---

9 ADF, *Military Ethics*, 36.
10 Paolo Tripodi, 'Understanding Atrocities: What Commanders Can Do to Prevent Them' in *Ethics, Law and Military Operations*, edited by David Whetham (London: Palgrave, 2010), 179.
11 Although this book states that there are 6 individual common cognitive behaviours, there are in fact 7 theories due to the counting of social comparison theory and conformity as one.

evaluation being based on comparison with other persons.'[12] Basically people use the opinion of others as a basis to judge their own abilities, attitudes and judgements. Social comparison theory has subsequently been defined as a behaviour in which individuals compare 'behaviours and opinions with those of others in order to establish the correct or socially approved way of thinking and behaving.'[13] Conformity was first discussed by Solomon Asch in 1951 in the paper *Effects of Group Pressure upon the Modification and Distortion of Judgments*. In this paper Asch found that 'a substantial minority yielded, modifying their judgments in accordance with the majority.'[14] This research demonstrated how over 70 per cent of those taking part in the experiment followed the groups irrational lead, rather than make their own more rational decision. Conformity is defined as a change in 'deep-seated private and enduring behaviour and attitudes due to group pressure.'[15]

There is a strong social obligation to conform to the social norm and research has demonstrated that this 'personal consistency is highly valued in our culture ... we easily fall into the habit of automatically being so.'[16] This tendency to fit into the social norm is even stronger in cohesive organisations such as the military. In these settings these theories are often called peer pressure, where soldiers feel that they must act in a certain way because their peers either want or expect them to. The International Committee of the Red Cross (ICRC) research confirmed that 'fighting men are generally motivated more by group pressure than by hatred or even fear.'[17] The military exploits this social pressure to ensure that units act efficiency, however when in stressful situations, enhanced by situational influencers this obligation to follow the social norm and conform can lead soldiers to conduct unethical actions in order to fit in to the group. Two examples will now be explored; the social comparison theory case study will explore the Australian Special Air Service (SAS) killings in Afghanistan whilst the conformity case study will explore the My Lai massacre in Vietnam.

---

12 Leon Festinger, 'A Theory of Social Comparison Processes', *Human Relations 1954*, Vol 7 (1954), 140.
13 Michael Hogg and Graham Vaughan, *Social Psychology, 7th Ed.* (Harlow: Pearson, 2014), 664.
14 S. E. Asch, 'Effects of Group Pressure upon the Modification and Distortion of Judgement', in *Group. Leadership and Men*, edited by H. Guetzkow (Pittsburgh: Carnegie Press, 1951), 190.
15 Hogg and Vaughan, *Social Psychology*, 658.
16 Robert Cialdini, *Influence: The Psychology of Persuasion.* (New York: Harper Collins, 2007), 60.
17 ICRC, *The Roots of Behaviour in War: Understanding and Preventing IHL Violations.* (Geneva: ICRC, 2004), 6.

Social Comparison Theory Case Study – Australian Special Air Service Killings

Operation Slipper was the longest continuous deployment for members of the Australian Defence Force (ADF) in Afghanistan. The Special Operations Task Group, mainly consisting of the Australian SAS, was key to this operation with many deploying six or more times. Major General Paul Le Gay Brereton investigated killings committed by the Australian SAS between 2009 and 2013. He produced *The Brereton Report* which found evidence of the murder of 39 civilians and prisoners by 25 SAS personnel which were subsequently covered up by the ADF. The report sited situations in which junior soldiers were required by their leaders to murder non-combatants to get their first kill. This practice became commonly known as blooding. They would then place a weapon by the prisoner and take a photograph as evidence that the prisoner was a combatant and therefore legitimise the killing.

The Brereton Report explained how to have a kill and treat civilians with no respect became the social norm. It stated that there was 'a normalisation over time of behaviours that should never have been considered normal.'[18] It also noted that there was 'the cultural normalisation of deviance from professional standards.'[19] The lower levels of leadership within the Australian SAS had developed their own set of social norms and all those new to the organisation conformed to the leaders wishes. The report emphasised how 'ethical leadership was compromised by its toleration, acceptance and participation in a widespread disregard for behavioural norms.'[20] In this environment the routine abuse of prisoners and killing of civilians became acceptable and the report concluded that 'this demonstrates how behaviour becomes embedded at the level of organisational culture, which then determines what is considered 'normal".'[21] In this operation, the power of the situation combined with a lack of supervision at the lower levels allowed for an unethical social norm to develop and this was combined with the strong ethos to conform, which ultimately resulted in the deaths of at least 39 innocent civilians.

---

18 ADF, *Inspector-General of The Australian Defence Force Afghanistan Inquiry Report, Part 1 – The Inquiry, Part 3 – Operational, Organisation and Cultural Issues.* (Canberra: ADF, 2020), 504.
19 Ibid., 122.
20 Ibid., 395.
21 Ibid., 518.

Conformity Case Study – My Lai Massacre

The My Lai massacre was the murder of up to 500 unarmed Vietnamese civilians by C Company, 1st Battalion, 20th Infantry Regiment, US Army in South Vietnam on 16 March 1968. The company had been told by its commander, Captain Ernest Medina to expect a hard fight when they arrived in My Lai. One of the team leaders, Sergeant Hodge, remembered Captain Medina's orders and that 'it was quite clear that no one was to be spared in the village.'[22] When the US troops arrived there were in fact just old men, women and children preparing breakfast. The murders and abuse began with a soldier from the 1st Platoon who pushed a villager down a well and threw a grenade after them. The mass killings then started with one of the platoon commanders, Lieutenant William Calley, leading the way. Calley's Platoon put a group of around 80 civilians into an irrigation ditch and opened fire killing most of them. Paul Meadlo, a rifleman who helped gather a group of civilians into the centre of the village recalled that Calley came whilst he was guarding the group and said '"I want them dead." So he started shooting them. And he told me to start shooting. I poured about four clips into them.'[23] As the killings began to expand mutilations and rapes followed. Private Varnado Simpson remembered that 'you didn't have to look for people to kill; they were just there. I cut their throats, cut off their hands, cut out their tongue, their hair, scalped them. I did it. A lot of people were doing it, and I just followed.'[24] A US Army photographer, Ronald Haeberle, recalled that he 'noticed a woman [who] appeared from some cover and this one GI fired first at her, then they all started shooting at her, aiming at her head. The bones were flying in the air chip by chip.'[25] Private Michael Bernhardt commented on the attitude of the group when he stated that 'the definitions for things were turned around. Courage was seen as stupidity. Cowardice was cunning and wariness, and cruelty and barbarity were seen sometimes as heroic.'[26]

The My Lai massacre highlights the strength of conformity amongst US troops in Vietnam, which was considered an extremely hostile environment

---

22 Tony Raimondo, *The My Lai Massacre: A Case Study*. School of the Americas, Fort Benning. https://downloads.paperlessarchives.com/p/Eyjx/#:~:text=The%20My%20Lai%20Massacre%3A%20A,the%20school's%20human%20rights%20program.
23 Lindsey Drew, 'Something Dark and Bloody', *The Quarterly Journal of Military History*, Vol 25, No. 1 (2012), 56.
24 Ibid., 55.
25 Ibid., 54.
26 Ibid.

where troops were constantly exposed to normalise violence. A US Marine Corps Lieutenant in Vietnam, Philip Caputo, who later went on to be Pulitzer Prize-winning journalist, confirmed that 'our mission was not to win terrain or seize positions, but simply to kill: to kill Communists and kill as many of them as possible.'[27] A journalist at the time of the massacre substantiated this noting 'killing civilians and destroying their villages had come to be the rule, and not the exception.'[28] The social norm of killing civilians and the peer pressure to conform to the social obligation to fit in had resulted in the situation in which the deaths of around 500 innocent civilians was seen to be morally acceptable by those present. This acceptance of mass killing was further reinforced when General Westmoreland, the commander of US armed forces in Vietnam, on hearing of the high body count congratulated the unit for doing a good job, rather than considering that unethical action might have taken place.

Deindividuation

Deindividuation was first discussed by Leon Festinger, Albert Pepitone and Theodore Newcomb in 1952.[29] Deindividuation refers to the 'loss of one's sense of individuality during which the person behaves with little or no reference to personal internal values or standards of conduct.'[30] It has also been defined as the 'process whereby people lose their sense of socialised individual identity and engage in unsocialised, often antisocial, behaviours.'[31] Deindividuation is similar, but not the same as conformity. Conformity refers to a person's conscious decision to adjust to the norms of a group; but deindividuation takes this a step further by considering the concepts of anonymity, loss of self-consciousness and the outsourcing of moral thinking. In this state, when people feel anonymous, they don't believe that they are responsible for their individual actions, but that the group or the leader hold the responsibility. Further research identified 'anonymity as one of the main factors that influences deindividuation.'[32] Therefore in a stressful situation,

---

27 Ibid.' 57.
28 Raimondo, 'The My Lai Massacre: A Case Study'
29 L. Festinger, A. Pepitone and T. Newcomb, 'Some consequences of deindividuation in a Group', *Journal of Social Psychology*, Vol 47 (1952).
30 J. Roeckelein, *Elsevier's Dictionary of Psychological Theories*. (Fountain Hills: Elsevier, 2006), 2004.
31 Hogg and Vaughan, *Social Psychology*, 658.
32 Jenna Chang, 'The Role of Anonymity in Deindividuated Behaviour: A Comparison of Deindividuation Theory and the Social Identity Model of Deindividuation Effects (SIDE)', *The Pulse*, Vol 6, No. 1 (2008), 6.

when there is a strong group ethos or domineering leader, individuals can lose their moral identity to the group and engage in unethical actions with the crowd that they would not conduct if they were alone.

Deindividuation Case Study – Amritsar (Jallianwala Bagh) Massacre

In April 1919 Brigadier-General Reginald Dyer gave orders for troops from the British Indian Army to open fire on a gathering of protesters in the Jallianwala Bagh in Amritsar, killing over 300 Indian civilians. A few days prior to the massacre there had been riots and deaths of both Indian and European residents in the city. Brigadier Dyer had been sent to the city to restore order. He had made a proclamation outlawing gatherings on the 13 April but was told that around 1,000 Indians had gathered in the Jallianwala Bagh, an enclosed square in the city. He immediately set of with 90 troops, 50 riflemen, 25 from the Frontier Force (54th and 59th Scinde Rifles), 25 from the 1st Battalion, 9th Gurkha Rifles and 40 Gurkha troops who were not armed. He stood down the troops company commanders and appointed Captain Crampton of the Gurkhas to command all the troops. It is believed he did this as 'by dismissing the middle-ranking officers, Dyer ensured that there would be no officers present who might bulk at his plans.'[33] On arriving at the narrow entrance to the square he was faced with a crowd of 25,000 Indians. Dyer immediately shouted 'Gurkhas right, 59th left. Fire.'[34] Witnesses state that 'the order was repeated by Captain Compton, whistles rang out, and immediately the troops opened fire.'[35] It was also noted that 'the men obeyed him implicitly.'[36] He later said to the Hunter Inquiry into the massacre that 'I realised that my force was small and to hesitate might induce attack. I immediately opened fire and dispersed the crowd.'[37] He later told the Inquiry that 'I had made up my mind that if I fired I must fire well and strong so that it would have a full effect. I had decided if I fired one round I must shoot a lot of rounds or I must not shoot at all.'[38] The shooting killed hundreds and went on for ten minutes, while hundreds more crushed each other to death trying to escape.

---

[33] Nigel Collett, *The Butcher of Amritsar: General Reginald Dyer.* (London: Cambridge University Press, 2005), 257.
[34] Alfred Draper, *Amritsar: The Massacre that ended the Raj.* (London: Cassell, 1981), 87.
[35] Nigel Collett, *The Butcher of Amritsar: General Reginald Dyer.* (London: Cambridge University Press, 2005), 259.
[36] Draper, *Amritsar,* 87.
[37] Nick Lloyd, 'The Errors of Amritsar', *BBC History Magazine,* Vol 10, No. 2 (2009), 53.
[38] Ibid., 53.

The British Indian troops who accompanied Dyer immediately followed the orders and continued to fire even when they could see the destruction that was unfolding. Sergeant Wiliam Anderson, a British witness to the massacre stated that 'the fire control and discipline of the native troops as first class.'[39] It was also noted that 'the soldiers were as disciplined and calm as marksmen at the butts, and there was no wild or sporadic firing ... Briggs [a British captain] was most impressed that not one man had hesitated and had fired high.'[40] The troops therefore not only immediately started firing, but did so in an extremely efficient manner. In an interview with two of the Gurkha soldiers after the event they were 'asked what they thought of the incident they both replied with evident relish: "Sahib, while it lasted it was splendid: we fired every round we had".'[41] There only appeared to be one person on the British side that was not suffering from deindividuation and that was Dyer's aide, Captain Briggs. Sergeant Anderson recorded that 'after a bit, I noticed that Captain Briggs was drawing up his face as if in pain', so Briggs was obviously distressed. Anderson went on that Brigg's began 'plucking at the General's elbow.'[42] There is unfortunately no more information about what happened next apart from that Dyer ignored the aide. However, it does highlight that this young officer considered that the action was not morally right. Nearly all the writing and research into this massacre is focused on Dyer and there is almost nothing about the young officers or 50 riflemen who carried out the massacre, they were anonymous and just following the orders of an overbearing leader. This is classic deindividuation; they simply did as they were told in an exceptionally efficient manner without feeling responsible for their actions. They considered themselves and remain today simply a tool used to execute Dyer's extreme violence.

Obedience

Obedience was first discussed by Stanley Milgram in 1963 following an experiment about authority. He believed that 'obedience is the psychological mechanism that links individual action to political purpose. It is the

---

39 Kim Wagner, *Amritsar 1919: An Empire of Fear and Making of a Massacre*. (London: Yale University Press, 2019), 166.
40 Draper. *Amritsar*, 88.
41 Wagner. *Amritsar 1919*, 172.
42 Ibid., 165.

dispositional cement that binds men to systems of authority.'[43] Like many in America, Milgram had been affected by the Eichman Trial and Hannah Arendt's concept of the banality of evil. He wondered if Americans could be induced to commit acts against their conscience as the Germans had done under the Nazis. The experiment and the application of shocks by the teacher has already been outlined in chapter 1.. The main reason for people giving these shocks was that they were obeying the person in the white coat who was facilitating the experiment and someone whom they considered to be an authority figure. The authority figure, an actor called Mr Williams, had four prompts that were used to enforce obedience. They got more directive each time so went from 'Please continue' to 'The experiment requires you to continue' to 'It is absolutely essential that you continue' and finally 'You have no other choice but to continue.'[44] If the teacher did not obey these prompts, the authority figure had two extra prompts that could be used these were 'although the shocks may be painful, there is no permanent tissue damage, so please go on' and 'whether the learner likes it or not, you must go on until he has learned all the word pairs correctly. So please go on.'[45] However, most teachers obeyed without the need to use this extra encouragement. Milgram believed that people obeyed these orders against their better judgement for a number of reasons, including the formality of the situation, the influence of an authority figure and the supposed legitimacy of the experiment.

Obedience to authority is ingrained in us all from an early age by parents and at school. A recent definition of obedience is that it is a 'behaviour in compliance with a direct command, often one issued by a person in a position of authority.'[46] In the military obedience is expected, personnel are trained from their first encounter with leaders to follow orders instantly so that on the battlefield there is no hesitation. However, once again the very action that the military uses to make it efficient in combat can have devastating effects if soldiers obey orders that they morally know to be wrong but under the power of the situation feel compelled to follow and result in unethical actions. Mark Osiel suggests that leaders should be accountable and punished not only for manifestly illegal acts such as massacres and atrocities, but 'for any

---

[43] Stanley Milgram, 'Behavioural Study of Obedience', *Journal of Abnormal and Social Psychology*, Vol 67 (1961), 371.
[44] Stanley Milgram, *Obedience to Authority: An Experimental View*. (New York: Harper Collins, 1974), 21.
[45] Ibid., 22.
[46] APA Dictionary of Psychology. https://dictionary.apa.org/obedience.

crimes resulting from an unreasonably mistaken belief that a superior's orders were lawful.'[47]

Obedience Case Study – Hola Camp Massacre

Hola camp was established to house Mau Mau detainees classified as hardcore by the British establishment during the Kenyan insurgency. By January 1959 the camp had a population of 506 detainees of whom 127 were held in a special remote camp.

This remote camp was reserved for the most uncooperative of the detainees who had refused to work on an irrigation system. To counter this the British Assistant Commissioner of Prisons John Cowan and the Camp Commandant Michael Sullivan drew up a plan of action to be adopted to force these reluctant detainees to work. This plan became known as *The Cowan Plan*. *The Cowan Plan* relied on prison guards using force to get the detainees to work.

On 3 March 1959, he put this plan into action; Sullivan 'briefed his warders that, in the event of the detainees creating trouble, batons might be used, blows being aimed at the legs of the detainees.'[48] This was against the prisons standing orders which stated weapons of force were not be used against prisoners 'except in self-defence or in defence of another person a prison officer must never strike a prisoner, but in order to overcome violence or resistance to escort he may use force but no more than is absolutely necessary.'[49] On the day of the massacre 88 detainees refused to work so the 'guards threw the prisoners into a ditch' which was four metres deep.[50] The trench was surrounded by guards and 30 guards got into the trench with the detainees. Cowan then blew a whistle and on this they obediently 'beat the defenseless [sic] men until the soldiers "were too tired to continue".'[51] The beatings went on for three hours, after which 11 men were dead and the other 77 were maimed or wounded. In this incident the guards gave up feelings of

---

47 Mark Osiel, *Obeying Orders: Atrocity, Military Discipline and the Law of War*. (London: Routledge, 2017), 945.
48 Parliamentary Archives: British Parliamentary Papers, House of Commons Debate: Hola Detention Camp, Volume 607, 16 June 1959. https://hansard.parliament.uk/Commons/1959-06-16/debates/b8d721fb-856c-4c71-8ecf-b05f477837dd/HolaDetentionCamp.
49 Ibid.
50 Joshua Scullin, *The Mau Mau Insurrection: The Failed Rebellion That Freed Kenya*. Undergraduate Thesis, University of Washington Tacoma, 2017, 18.
51 Ibid.

personal responsibility to blindly obey the camp commandant's instructions regardless of the damage they were doing to the detainees.

Cognitive Dissonance Theory

Cognitive dissonance theory was first discussed by Leon Festinger in 1957.[52] He explained that 'if a person knows various things that are not psychologically consistent with one another, he will, in a variety of ways, try to make them more consistent.'[53] It has also been defined as a 'state of psychological tension produced by simultaneously having two opposing cognitions.'[54] When a person has cognitive dissonance they are generally behaving in a way that they know to be wrong, but conducting an activity which they want to do creating friction between actions and beliefs. In this situation Festinger proposed that we try to seek harmony in our beliefs and behaviours and try to reduce tension that this inconsistency causes. This harmony is normally produced by justifying the actions, for example by blaming the victim. Abert Bandura's work on moral disengagement equated these ideas to the attribution of blame. Bandura noted that in this emotional state of cognitive conflict people often choose to do what is wrong and then mitigate their actions in an attempt to reconcile their behaviour with their beliefs. It has sometimes been associated with the idea of having an angel and a devil on our shoulder telling us what to do, we know what is right, but we want to do something else that we know to be wrong.

In 1970, John Darley and Daniel Batson ran an experiment at the Princetown Theological Seminary to test this theory. They informed 40 students that they would be giving a presentation, half of them on the parable of the Good Samaritan. Some students were told that they had plenty of time to get to the room and deliver the talk, some were told that they needed to get there quickly and a third group were told that they were late and needed to get there immediately. However, on the way they passed down a narrow alley which had a figure slumped on the ground and moaning (in a similar way to the parable of the Good Samaritan). Sixteen students or 40 per cent helped to varying degrees, whilst 24 or 60 per cent hurried by. The figures were not substantially affected by what the students were going to teach but affected by the time they believed they had. Of those not in a rush 63 per cent helped, of

---

[52] L. Festinger, *A Theory of Cognitive Dissonance* (Stanford: Stanford University Press), 1957.
[53] Leon Festinger, 'Cognitive Dissonance', *Scientific American,* Vol 207, No. 4 (1962), 93.
[54] Hogg and Vaughan, *Social Psychology,* 657.

those who were in a medium rush 45 per cent stopped, but of those told that they were late only 10 per cent helped. Darley and Batson commented on these students that 'ironically, he is likely to keep going even if he is hurrying to speak about the parable of the Good Samaritan, thus inadvertently confirming the point of the parable.'[55] Their conclusion was that 'whether a person helps or not is an instant decision likely to be situationally controlled.'[56]

In stressful military situations personnel are often faced with the choice between doing what they ought to do or doing what they want to do. There are many possible scenarios, from giving medical treatment to an injured enemy fighter, abusing prisoners or detainees for fun, to situations already examined such as the murders by the Australian SAS or the abuse and killing at My Lai. Often people are aware of the right thing to do, but as examined due to situational pressure they make an instant decision not to do the right thing but to do the easy or enjoyable thing. They then have to justify or mitigate their action and this is often done by dehumanising or demonising their victims (which will be explored in the next chapter). In some cases, personnel decide to cover up their unethical decision by making even more extreme and riskier decisions, such as executing witnesses. In the Batang Kali massacre case study this extreme cognitive dissonance is explored.

Cognitive Dissonance Theory Case Study – Batang Kali Massacre

On the 12 December 1948 a patrol from the 2nd Battalion Scots Guards murdered 24 male workers on a rubber plantation in Batang Kali during the early stages of the Malaya Emergency. The patrol of 14 soldiers from 7 Platoon, G Company was commanded by Sergeant Charles Douglas. They had arrived in the small workers hamlet on the Sungei Remok Estate on 11 December and were accompanied by a Chinese Special Branch officer, a Malay and a Tamil special constable. When they entered the village they questioned a youth named Loh Kit Lin. Whilst this was happening, Sergeant Douglas shot Lin dead and just after an elderly villager, Chan Loi, had a heart attack. As night began to fall all the villagers were imprisoned overnight. In the morning Sergeant Douglas split the 25 men into four groups and these were executed at close range in the village. There was one survivor, Chong Fong, who fainted and was believed to be dead. When the patrol

---

[55] John Darley and Daniel Batson, 'From Jerusalem to Jericho: a Study of Situational and Dispositional Variables in helping Behavior', *The Research Experience, Experiment* (1973), 202.
[56] Ibid., 204.

returned to camp they informed their commanders that they had ambushed 25 bandits (the name for communist insurgents in 1948) who were trying to escape. The British Army had been on the back foot for the first six months of the emergency and this was treated as great news. Soon the success of the elite Scots Guards was being sent around the World. The Singapore based *Straits Times* reported it as the 'Biggest Success for Forces since the Emergency Started.'[57] But this lie did not hold up for long as 'by 22 December Thomas ('Tom') Menzies, European owner of the rubber estate, had claimed the victims were ordinary workers.'[58] Even after several investigations, no member of the patrol was brought to justice.

It has not been established why Sergeant Dougles shot Loh Kit Lin on the 11 December, however once this action had taken place, he decided to cover up shooting this individual by killing all the males in the village and fabricating the ambush story. Sergeant Douglas knew he had done wrong, the Malay special constable that had accompanied the patrol later said that 'the sergeant warned me that if I breathed a word to anyone about the shooting I would land myself in jail.'[59] The 12 Guardsman who took part in the killings also knew that it was wrong. One of those that took part, Guardsman William Cootes, later stated that 'I did not want to kill anybody, but was too frightened to move and make myself look a coward in front of the others.'[60] Guardsman Robert Brownrigg recalled that 'this was needless killing that was like murder under orders. There was no need to kill these people.'[61] Another member of the patrol, Guardsman Alan Tuppen, recalled that 'once the firing started we seemed to go mad…. Some of the men were excited, some were delighted, some of us stayed quiet. It struck me we must all be out of our minds to do a thing like we had just done.'[62] Here there is a good demonstration of cognitive dissonance as Guardsman Tuppen's actions and beliefs are in conflict. He knew as did the others that executing the male villagers was wrong, but under the orders of Sergeant Douglas and a sense of group cohesion, the Guardsman killed the villagers and later regretted their actions made under the power of the situation.

---

57 Christpher Hale, 'Batang Kali: Britain's My Lai', *History Today* (2012), 4.
58 Karl Hack, 'Devils that suck the blood of the Malayan People: The Case for Post-Revisionist Analysis of Counter-insurgency Violence', *War in History*. Vol 25, No. 2 (2018), 209.
59 Anthony Short, 'The Malayan Emergency and the Batang Kalii Incident', *Asian Affairs,* Vol 41, No. 3 (2010), 345.
60 Ian Ward, and Norma Miraflor. *Slaughter and Deception at Batang Kali,* Singapore: Media Masters, 2009, 88.
61 Ibid., 89.
62 Ibid., 88.

Bystander Effect

The bystander effect was first discussed by John Darley and Bibb Latane in 1968 in their paper *Bystander Intervention In Emergencies: Diffusion of Responsibility*. In this they believed that 'the more bystanders to an emergency, the less likely, or the more slowly, any one bystander will intervene to provide aid.'[63] The bystander effect has also been defined as the concept where 'people are less likely to help in an emergency when they are with others than when alone. The greater the number, the less likely it is that anyone will help.'[64] The second part of Darley and Latane's work was around the concept of the diffusion of responsibility. The diffusion of responsibility is the 'tendency of an individual to assume that others will take responsibility.'[65] Albert Bandura's moral disengagement work describes how in a situation 'where everyone is responsible, no one feels personally responsible.'[66] The UK MOD has done some work on the diffusion of responsibility and how as a group increases in size the likelihood of people helping reduces. Their findings were that when '1 person witnesses a situation = 85% will help, 3 people witness a situation = 62% will help, 5 people witness a situation = 31% will help.'[67] These ideas reinforce the original work conducted by Darley and Latane, who in turn took their impetus for the study of a murder which took place in New York in 1964. The victim was a bar manager called Kitty Genovese, who arriving home was attacked and stabbed twice crying out for help. Her attacker ran away, but when no help came, he returned, stabbed her again killing her and stealing her bag, the whole incident had lasted around half an hour. The newspapers sensationalised the idea that 37 people had witnessed the attack, but nobody took responsibility believing that others would help or phone the police. The truth is that two people did call the police and one person shouted at the attacker from their apartment. Nevertheless, the media story was the catalyst for research which ultimately culminated in the codification of the bystander effect.

In a series of experiments, it has been shown that the more people present, the less likely they are to provide assistance as everyone believes that someone else will act. This diffusion of responsibility is also prevalent

---

63 J. Darley and B. Latane, 'Bystander Intervention in Emergencies: Diffusion of Responsibility', *Journal of Personality and Social Psychology*, Vol 8, No. 4 (1968), 378.
64 Hogg and Vaughan, *Social Psychology*, 657.
65 Ibid., 658.
66 Bandura, 'Moral Disengagement', 3.
67 MOD, *Active Bystander Fundamentals*, 2023.

The S-CALM Model

in the military, most leaders are generally comfortable with enforcing rules and procedures within their own chain of command. However, given an ambiguous situation some leaders prefer to turn the other way than tackle a situation. The classic example of this at the routine level is the saluting of officers. Soldiers know that they should do this, but many will not salute an officer if they do not know them and many officers who are not saluted by soldiers will not correct the behaviour. This may seem a small thing, but it is indicative of the behaviour that given the power of the situation and the spiral of violence can result in unethical actions.

Bystander Effect Case Study – Somalia Affair

The Somalia affair was a 1993 Canadian military scandal, prompted by the beating to death of Shidane Arone, a Somali teenager, at the hands of two soldiers from the Canadian Parachute Regiment. Shidane Arone had been detained by Captain Michael Sox at around 2045 hours on 16 March 1993 on the suspicion that he was trying to break into the Canadian camp and steal supplies. Captain Sox had given direction that morning that 'I don't care if you abuse them but I want those infiltrators captured.... abuse them if you have to. I do not want weapons used. I do not want gun fire.'[68] Captain Sox handed Shidane Arone to Warrant Officer Murphy who kicked Arone which was later taken to be implicit permission by Master Corporal Clayton Matchee and Trooper Kyle Brown to continue the abuse. Corporal Matchee and Brown took Shidane Arone to an ammunition bunker and started their abuse. Sergeant Mark Boland, a section commander, visited in the evening and saw Trooper Brown punch Shidane Arone in the jaw. As Sergeant Boland 'said to Pte Brown and MCpl Matchee, "I don't care what you do, just don't kill the guy".'[69] Later 'Matchee, intended to burn the soles of the Somali's feet with a cigarette. Sgt Boland reportedly said, "Don't do that, it would leave too many marks. Use a phone book on him".'[70] As the evening wore on and the abuse continued Corporal Matchee and Trooper Brown beat Shidane Arone until he died at around midnight.

---

68 George Shorey, 'Bystander Non-Intervention and the Somalia Incident', *Canadian Military Journal*, Winter 2000-2001 (2001), 24.
69 Canada. *Commission of Inquiry into the Deployment of Canadian Forces to Somalia and Donna Winslow. The Canadian Airborne Regiment in Somalia, a Socio-Cultural Inquiry: A Study Prepared for the Commission of Inquiry into the Deployment of Canadian Forces to Somalia*. Canadian Government Publishing, 1997, 321.
70 Ibid., 322.

It has been estimated that between 15 and 80 soldiers could hear the torture but did not intervene. Certainly, there were people in the immediate area who did not take action. For example, at the start of the abuse when 'Sgt Boland arrived shortly before 2100 hours to relieve MCpl Matchee. At that point, Maj Seward, Capt Sox, MWO Mills, and WO Murphy were in or around the bunker.'[71] It was also noted 'that during the course of the evening perhaps seventeen individuals came by the bunker, looked in, and left without commenting or interfering.'[72] In addition to these people close to the bunker it was also noted that 'during the time that Mr. Arone was being tortured and beaten to death, there were a number of Canadian soldiers in both the command and sentry posts.'[73] At one stage Corporal MacDonald working in the command post which was around 200 metres from the bunker 'reported hearing a "yelp" from the bunker'. He later testified that '"I recall everybody kind of looking in the direction of the bunker, and then just kind of went back to what they were doing".'[74] In this situation many of the Canadian soldiers were aware of the beating of Shidane Arone, but they took no action to stop it. There was a diffusion of responsibility with most soldiers believing that it was not their responsibility to step in as others, many of them higher in rank, must also have heard the torture but had not stepped in to stop it. This lack of confidence to do the right thing often results in unethical actions not being stopped by people who know that they are witnessing unacceptable behaviours.

Status Quo Bias

Status quo bias was first discussed by William Samuelson and Richard Zeckhauser in 1988. They defined status quo bias as 'doing nothing or maintaining one's current or previous decision.'[75] They concluded that 'in choosing among alternatives individuals display a bias toward sticking with the status quo.'[76] Individuals tend to accept the current situation rather than take a risk on an outcome of change which might be uncertain. This preference

---

71 Ibid., 321.
72 Shorey, 'Bystander Non-Intervention and the Somalia Incident', 21.
73 Ibid.
74 Canada, *Commission of Inquiry into the Deployment of Canadian Forces to Somalia and Donna Winslow*, 324.
75 W. Samuelson and R. Zeckhauser, "Status Quo Bias in Decision Making". *Journal of Risk and Uncertainty*, Vol 1 (1988), 8.
76 Ibid., 47.

for the status quo and stability can cloud the decision-making process, with more weight being given to current circumstances. It can lead some leaders to display an unwillingness to drive change in their organisation and stick to what is already established and has worked before.

The military is traditionally a conservative environment. It can accept slow, evolutionary change, but can resist quick, revolutionary change. People tend to put up three barriers to change capability, motivational and opportunity barriers. The first are capability barriers, which focus on physical and psychological issues such as physical ability, cognitive skills, interpersonal skills and attention span. The second are motivational barriers, which are both automatic and reflective, these barriers include belief in abilities, beliefs about the consequences, negative emotions, unhelpful habits and accountability. Finally, there are opportunity barriers which focus on social and physical opportunities. These include issues such as opportunities in the environment, resources and time, social and cultural norms. In addition, in the military the use of status quo bias in units can often be disguised as tradition. The British Army has many traditions, and the majority are healthy and maintain a units *spirit de corps*. However, research has shown that 'when poor or undesirable behaviours are institutionalised, they are embedded in organisational memory, solidified in routines and structures' in these situations traditions can lead soldiers to conduct unethical actions, such as initiation ceremonies or abuse.[77]

Status Quo Bias Case Study – Deepcut Deaths

The Deepcut deaths took place at 25 Training Support Regiment, at the Princess Royal Barracks Deepcut between 1995 and 2002. It involved the deaths in obscure circumstances of four British Army trainee soldiers, Privates Benton, James, Gray and Collinson. On 9 June 1995, Private Sean Benton died from five bullet wounds in his chest after going on an unauthorized lone patrol of the camp perimeter. On 27 November 1995, Private Cheryl James died from a single bullet wound in her head, her body was found in a wood not far from her abandoned guard post. On 17 September 2001, Private Geoff Gray died from two gunshot wounds to the head whilst on guard duty, having left colleagues to carry out a lone prowler patrol in contravention of the routine procedure. Finally on 23 March 2002, Private James Collinson

---

[77] Samantha Crompvoets, *Blood Lust, Trust and Blame*. (Victoria: Monash University Publishing, 2021), 43.

died from a single gunshot wound to the head whilst performing guard duty: Collinson's body being found near the perimeter fence.

In December 2004 a military lawyer Queen's Council, Nicholas Blake, was commissioned by the MOD to independently review the cases with the voluntary co-operation of involved parties. In March 2006 he published a 397-page report entitled *The Deepcut Review* which pointed to complacency over a number of years which resulted in the deaths of the four trainees. On the lack of NCOs he noted that a 'more compelling argument to persuade the higher echelons of the Army to support these proposals [for better supervision] would have been an assessment of the nature of the risk of harm to the recruits and trainees.'[78] He went on to assess that 'if the problem of poor supervisory ratios in training regiments could not be immediately dealt with, other aspects of the overall problem needed to be addressed, and a clear understanding of what the Army faced by way of risk if the status quo remained.'[79] He also noted that an Army internal report in December 2002 'concluded that the status quo with regard to both supervisory ratios and trainees performing guard duty was unacceptable as a matter of risk.'[80] The Army establishment was content with the status quo at Deepcut and although this may not have stopped the two deaths in 1995, it may have contributed to the later deaths. In the Government's response to the Blake report they stated that 'there were failures to identify potential sources of risk and to address them.'[81]

---

[78] Nicholas Blake QC, *The Deepcut Review: A Review of The Circumstances Surrounding The Deaths of Four Soldiers at Princess Royal Barracks, Deepcut Between 1995 And 2002*. (London: The Stationery. 2006), 298.
[79] Ibid., 304.
[80] Ibid., 372.
[81] MOD, *The Government's Response to the Deepcut Review*. (Crown Copyright, 2006), 2.

4

Common Group Behaviours

Introduction

In 2004, a year after the invasion of Iraq, the US Government realised that they had won the war but lost the peace. George W. Bush asked Tony Blair for a British Army team to deploy and teach the incoming US forces the skills learnt during the British Army's 30 year counter insurgency campaign in Northern Ireland. I was selected to be the Executive Officer for what was called the British Army Counter Insurgency Short Term Training Team (STTT) which deployed to Iraq to teach the incoming US Army III Corps British Counter Insurgency procedures. The STTT was based in Camp Victory in Baghdad. One of the training courses was to the unit that was responsible for guarding Camp Victory and during discussions with the Commanding Officer, he informed me that the unit had had 12 soldiers killed in the guard towers around the camp since they had deployed in Iraq. The whole unit seemed to consider this loss of people acceptable. I asked if I could help and on inspection of one of the guard towers the solution to their problem was quickly obvious. The towers had no backdrop and therefore the soldiers in the towers were outlined against the sky like a target. I suggested that the unit install camouflage nets as soon as possible to stop the soldiers being silhouetted. Immediately after the nets were put up the deaths and casualties ceased. This unit had been suffering from a series of group behaviours including groupthink, risky shift and status quo bias. These behaviours had led to unethical decisions in the unit and the unnecessary death of 12 of its soldiers.

Following on from the last chapter which investigated the common cognitive behaviours which tend to be identified in individuals, this chapter will explore the second group of six cognitive behaviours which mainly appear in groups or teams. As with the first six, they are not in themselves unethical, but under the power of the situation they can lead to unethical actions taking place. The six behaviours and case studies are: groupthink applied to the Nisour Square massacre in Iraq in 2007, risky shift demonstrated by the Nazis Reserve Police Battalion 101 in the Second World War, authority bias applied to the Chuka massacre in Kenya in 1953, othering as demonstrated by the My Li massacre in Vietnam in 1968, dehumanisation as applied to the Abu Ghraib abuses in Iraq in 2004 and demonisation as seen in the Voćin massacre in Croatia in 1991.

Groupthink

Irving Janis first discussed groupthink in the 1972 paper *Victims of Groupthink: A Psychological Study of Foreign Policy Decisions and Fiascos*. He believed that Groupthink was 'a deterioration of mental efficiency, reality testing, and moral judgment that results from ingroup pressures.'[1] In more common language, Groupthink has been defined as 'a sheeplike tendency to pick options that avoid arguments rather than get the job done.'[2] Groupthink has also been defined as 'a mode of thinking in highly cohesive groups in which the desire to reach a unanimous agreement overrides the motivation to adopt proper rational decision-making procedures.'[3] The idea of a group as a cohesive organisation is important for groupthink to function. In close knit teams 'when individuals categorize themselves as members of a group, they regard themselves as relatively interchangeable exemplars of the group rather than as unique individuals' therefore even when not in the group they can still carry its values.[4] Janis believed that there are eight symptoms of groupthink that can be identified.

The first symptom of groupthink is a feeling of invulnerability. Groups create excessive optimism and pay little attention to the consequences of their actions. The next symptom is rationalisation, in this the group discounts information that threatens the common belief. It also ignores or belittles information from outside of the group. The third symptom is group morality, in which the group believes that it acts only in a moral or ethical way. In the fourth symptom the group holds to stereotypes. Opponents to the group are

---

[1] I. Janis, *Groupthink: Psychological Studies of Policy Decisions and Fiascos*, 2nd ed. (Boston: Houghton Mifflin), 1972, 9.
[2] Murray, *Brains & Bullets*, 207.
[3] Hogg and Vaughan, *Social Psychology*, 660.
[4] Daniel Miller, Eliott Smith and Diane Mackie 'Effects of Intergroup Contact and Political Predispositions on Prejudice: Role of Intergroup Emotions'. *Group Processes & Intergroup Relations* 7(3), 2004, 222.

put into stereotypes and generally their abilities are underestimated by the group. The next symptom is group pressure. With this symptom dissenters have their loyalty questioned and are put under peer pressure to conform. The sixth symptom is that of self-censorship. This sees group members holding back from expressing any doubts about the group's decisions. The penultimate symptom is unanimity. This brings together some of the other symptoms and a lack of dissent or silence gives the illusion of consent during decision-making. The final symptom is concerned with mind-guards. Group members try to protect the group by restricting information that might go against the supposed consensus. Janis felt that these symptoms were 'a mutual effort by group members to maintain self-esteem and emotional equanimity.'[5]

The military relies on groupthink to ensure that all elements act in the same way during combat. The military is susceptible to groupthink as it is a highly cohesive, hierarchical group which operates in stressful situations. Janis was aware of this and noted that 'the more amiability and *esprit de corps* there is among the members of a policy-making in-group, the greater the danger that independent critical thinking will be replaced by groupthink.'[6] The ICRC have conducted research into the behaviour of groupthink amongst military units. Their first point was that 'the individual may not normally be a killer but the group certainly is. Many studies have shown that fighting men are generally motivated more by group pressure than by hatred or even fear.'[7] This notion that individuals are influenced by their peers was fully highlighted in the last chapter. The ICRC also noted that a group tends to generate 'prejudices, simplifications and discrimination' and 'when another group is declared to be an enemy, these tendencies become even more acute.'[8] In this case groups tend to close ranks and conduct collective rationalisation in the inherent belief in the morality of their decisions. The military reliance on groupthink and its susceptibility to the symptoms of it means that when a group acts unethically it can have disastrous outcomes.

---

5 Irving Janis, 'Groupthink', *IEEE Engineering Management Review*, Vol. 36, No. 1 (2008), 88.
6 Ibid., 85.
7 ICRC, *The Roots of Behaviour in War: Understanding and Preventing IHL Violations*. (Geneva: ICRC, 2004), 6.
8 Ibid., 7.

Groupthink Case Study – Nisour Square Massacre

On 16th September 2007 a team from Blackwater Security Consulting, a private military company known as Raven 23, was escorting a US embassy convoy in Iraq. Raven 23 was ordered to Nisour Square to control traffic for a second convoy that was approaching from the south. The second convoy was bringing diplomats who had been evacuated from a meeting after a bomb went off near the compound where the meeting was taking place. As Raven 23 drew close to Nisour Square, a car driven by Ahmed Haithem Ahmed Al Rubia'y was approaching the square, driving slowly on the wrong side of the road. The car ignored a police officer's whistle to clear a path for the convoy. One of the Blackwater staff, Nicholas Slatten, fired warning shots and then lethal fire at the car killing Ahmed at which time control of the car was lost and it speed towards the convoy. A narrative of the event continues that:

> He began to shoot at the unstoppable vehicle, killing the mother, Mahasin. Shots continued to ring out in Nisour Square now from all four Blackwater convoys, so much so that the vehicle Ahmed and Mahasin were killed in exploded into flames. Once the shooting ceased, fifteen vehicles had been destroyed, seventeen Iraqi civilians killed, and over twenty civilians wounded. It was later discovered that not only was shooting coming from the Blackwater convoy that day, but from Blackwater's Little Bird helicopters as well.[9]

It appears that once Slatten started to fire the whole unit, including the supporting helicopters joined in; Ahmed's car alone had 40 bullet holes in it. One report noted that 'witnesses described a horrifying scene of indiscriminate shooting by the Blackwater guards. In all, as many as twenty eight Iraqis may have been killed.'[10] However, an investigation into the events reported that the attitude within Blackwater was such that this event was inevitable, it described that 'aggressive, fast driving through crowded roads and the possibility of using pre-emptive force inevitably translated into collateral damage well before Nisour Square.'[11] This report concluded

---

9 Marina Harden, 'Blackwater USA: The Success and Failures of the Worlds Most Powerful Mercenary Army in the War on Terror', *Pepperdine Policy Review 2017*, Vol. 9 (2017), 33-34.
10 Jeremy Scahill, 'Making a Killing', *Nation 10/15/2007*, Vol. 285, No. 11 (2007), 21.
11 Eugenio Cusumano, 'Diplomatic Security for Hire: The Causes and Implications of Outsourcing Embassy Protection', *The Hague Journal of Diplomacy 12* (2017), 42.

that 'aggressiveness and the excessive use of lethal force did not just occur as isolated cases, but were a systemic problem.'[12] The Blackwater guards in Nisour Square and the helicopter pilots above opened fire and as a group did not consider their actions were extreme or unnecessary.

Risky Shift

Risky Shift was first discussed by James Stoner in 1961. In this paper he concluded that 'when the subjects reached decisions as members of a group, they tended to advocate significantly more risky courses of action than they had chosen when they reached decisions as individuals.'[13] He did further research and confirmed that 'unanimous group decisions were more risky than the average of the initial individual decisions.'[14] Until this time most people believed that groups were more cautious than individuals and that a group decision was ultimately safer. Stoner recognised this phenomenon and noted that 'subjects who agreed with the final group decision tended to become more confident of that decision.'[15] This behaviour is linked to many others, it relates to the individual behaviours of conformity and social comparison theory and the group behaviour of groupthink, in that individuals will go along with the majority opinion of the group. It also relies heavily on the diffusion of responsibility which leads to more extreme decisions being made. It is possible that the group could decide to take less risk, however it has been shown that 'arguments favouring risk taking are more powerful than those favouring cautiousness.'[16] Therefore people tend to be persuaded to go with the risker decision, other research has also discovered that 'groups are inherently less moral than individuals.'[17]

In a review of the fighting on Makin Island in 1943, *Men Against Fire*, Brigadier S. Marshall concluded that only 15 per cent of soldiers in defensive positions fired their weapons and almost all these were crew served. The contention was that soldiers in individual fox holes simply hid whilst those

---

12 Ibid., 43.
13 J. Stoner, *A Comparison of Individual and Group Decisions Including Risk*. MA Thesis, MIT, Boston, 1961, 65.
14 James Stoner, *Risky and Cautious Shifts in Group Decisions: The Influence of Widely Held Values*. Working Paper for Massachusetts Institute, Technology, Boston, 1967.
15 Stoner. *A Comparison of Individual and Group Decisions Including Risk*, 65.
16 Neil Malamuth and Seymour Feshbach, 'Risky Shift in a Naturalistic Setting', *Journal of Personality* (1972), 46.
17 King's College London, Centre for Military Ethics, Armouring Against Atrocity, 2016. https://militaryethics.uk/en/course/library.

serving together fought side by side. The British Army therefore now trains and fights in four-person teams. These teams build up trust and even if they take casualties the remainder of the team will fight on. Research has revealed that by '2009, the fire ratio for US, Canadian and British troops in Afghanistan looked to be touching 100 per cent. Almost every man is firing his weapon at some point during the fight.'[18] The idea that groups make riskier decisions than individuals is vital to the military, which is a risk-taking organisation, but in unethical situations it can lead to poor decision making.

Risky Shift Case Study – Police Battalion 101

Reserve Police Battalion 101 was raised in Hamburg and was initially used as a paramilitary force during the invasion of Poland in 1939 with the aim of guarding Polish prisoners of war. Following its initial duties, the Battalion was tasked in 1940 to evacuate Poles from the new German territories. In one action, they evicted 36,972 Poles, half of the overall target for the German Army. Later in the year, they were responsible for guarding the Jewish ghetto of Lodz for six months. This was the second largest ghetto with over 160,000 Jews crammed into a small urban space. They were then withdrawn to Hamburg and were given a new commander, Major Trapp, and training. This training was 'aimed at strengthening group bonds and ensure conformity to Nazi ideology.'[19] Following the offensive against the Soviet Union in June 1941, the first mass murder known to have been committed by Battalion 101 was the shooting of 1,500 Jews from the Jozefow ghetto in South Eastern Poland on 13 July 1942. For this, a generous supply of alcohol was provided. Before this initial action, 12 soldiers asked not to take part and were allowed to leave freely. Then on 17 August they executed 1,700 men, woman and children from the Lomazy ghetto. An analysis of this progression from killing men to killing everyone in a location noted that 'in July the women followed. From mid-August children considered as "useless eater" were included as a final "logical" step.'[20] Massacres and deportations continued throughout August to October. However, on 3 November 1942 the Battalion performed the largest single day massacre of the Holocaust, killing 43,000 Jews at Aktion Erntefest.

---

18 Murray, *Brains & Bullets*, 31.
19 Konrad Kweit, 'Hitler's willing executioners and "ordinary Germans" some comments on Goldhagen's ideas', Central European University, 1996.
20 Ibid.

This shift from no killings to small massacres of men to performing the single biggest massacre was a gradual shift. It was noted that one of the officers, Lieutenant Gnade, 'initially rushed his men back from Minsk to avoid being involved in killing but who later learned to enjoy it.'[21] There were also those that sort to ensure that everyone took part so that killing became the social norm. 'NCOs, like Hoppner and Ostmann, picked out individuals known as nonshooters and pressured them to kill, sometimes successfully.'[22] It was 'this process of brutalization and dehumanization' that helps to explain the shift in attitude and 'the ability and willingness to commit mass murder repeatedly' by the men.[23] As the months progressed they no longer needed excessive alcohol to commit these actions as 'volunteering for killing, as in many police battalions, was the battalion norm.'[24] In Christopher Browning's *Ordinary Men* he concludes that '80 to 90 percent of the men proceeded to kill, though almost all of them – at least initially – were horrified and disgusted by what they were doing'.[25] Due to risky shift, the battalion had pushed the boundaries of their collective behaviour to mass killing which they now considered to be legitimate.

Authority Bias

The effect of authority was first discussed by Stanley Milgram in 1963 as part of the Shock Experiment in which he noted that 'the individual who is commanded by a legitimate authority ordinarily obeys. Obedience comes easily and often.'[26] As earlier discussed, the idea of authority and obedience came from Hannah Arendt's comments on Adolf Eichmann. Milgram's detailed thinking on authority seem to stem from Carl Friedrich's 1958 book *Authority*. Milgram combined these ideas to suggest that authority is the tendency to attribute greater accuracy to the opinion of an authority figure and be more influenced by that opinion and that in this way it is similar and opposite to obedience. In this book authority and obedience have been investigated separately. Authority can be used without the prerequisite

---

21 Christopher Browning, *Ordinary Men: Reserve Police Battalion 101 and the Final Solution in Poland*. (New York: Harper Collins Books, 1980), 188.
22 Ibid., 171.
23 Kweit, 'Hitler's willing executioners'.
24 Daniel Goldhagen, *Hitler's Willing Executioners: Ordinary Germans and the Holocaust*. (London: Abacus, 1997), 252.
25 Browning, *Ordinary Men*, 183.
26 Stanley Milgram, 'Behavioural Study of Obedience', *Journal of Abnormal and Social Psychology*, Vol. 67 (1961), 371.

for obedience and this is termed authority bias. Authority bias has been defined as 'the tendency of people to blindly follow the advice, suggestions, or instructions of others who are in positions of authority.'[27] In this respect is similar to authority, however it also describes 'the tendency to believe or trust ideas that include some kind of badge or symbol of authority.'[28] This means that brands can have authority and in the military this can relate to units as well as badges of rank and qualifications.

Authority in the military is normally not questioned due to the hierarchy that exists in most armed forces. The Australian Defence Force doctrine confirms this idea stating that 'in hierarchical systems, the abuse of authority is an inherent risk.'[29] The power of an authority figure is magnified in the military where leaders have formal and legal trappings of authority, such as titles and wear their rank prominently as a mark of their power which influences the group. Research has shown that when military leaders display these legal trapping of authority 'their ability to engender atrocious behaviour is even more significant.'[30] The ICRC confirmed that 'ordinary men submit willingly to an authority when they believe that it is legitimate; they perceive themselves as its agents.'[31] Consequently a group can be dominated by a well-qualified and charismatic, but immoral leader, whom they may consider legitimate and will therefore conduct unethical actions on their orders.

Authority Bias Case Study – Chuka Massacre

During the Mau Mau insurgency in Kenya, B Company, 5th Battalion of the King's African Rifles (KAR) had been sent to the Chuka area on 13 June 1953, to flush out rebels suspected of hiding in the nearby forests. Although made up of African soldiers, the company had British officers and was commanded by Major Gerald Griffiths with Second Lieutenants Howard and Innes-Walker as platoon commanders. The operation in the Chuka area was typical of its type, the two platoons would conduct sweeps of the jungle while members of the local Home Guard policed the forest boundary.

---

27 Newristics. Authority Bias. Heuristic Encyclopaedia. https://newristics.com/heuristics-biases/authority-bias.
28 Ibid.
29 ADF, *Military Ethics*, 33.
30 King's College London, Centre for Military Ethics, Armouring Against Atrocity, 2016. https://militaryethics.uk/en/course/library.
31 ICRC, *The Roots of Behaviour in War*, 7.

Griffiths had as guides two ex-Mau Mau fighters, Njeru son of Ndwega and Kavenji son of Njoka, who had agreed to assist in finding the Mau Mau camps in the area.

Griffiths and the two junior British officers interrogated the guides. However, 'when the first prisoner seemed unwilling to cooperate, Griffiths ordered that a hole to made in his ear with a bayonet. A string was passed through the gaping wound, to be used as a tether over the next four days.'[32] The second guide was executed for being uncooperative and after a further four days of torture the first guide was also shot. On the 17 June, a patrol of ten men led by an African warrant officer moved out of the forest into the surrounding farmland to liaise with the Home Guard. It came across 12 members of the Home Guard gathered at a farmhouse. At sunset these men were executed. The following day a patrol with a British officer entered the village of Karege on the edge of the jungle. Again, it encountered a group of nine Home Guards and a child. This group was also executed and the hands of six of them were cut off as proof of the killings to claim a bounty for the liquidation of Mau Mau fighters. In addition, 'Griffiths himself admitted giving soldiers 5s 5d, [5 shillings and 5 pence] in contrast to others who offered a 5s reward for kills. He thought the practice perfectly normal.'[33] Griffith's disregard of local lives was emphasised during the subsequent court martial when Warrant Officer W. Llewellyn gave evidence about the briefing that Griffith's had given to B Company before the operation. In this he stated that 'Griffiths said: "You can shoot anybody you like – PWD [Public Works Department] or anybody." Asked what he understood Griffiths to mean, Llewellyn said: 'I understood we could shoot anyone black.'[34] When Second Lieutenant Innes-Walker was questioned as to why he was passive and did not try and stop the torture and killings he said, 'he was "in the habit of accepting his [Griffiths'] actions". This is an important statement in light of the debate on the fine line between the need for military obedience and the duty to refuse manifestly illegal orders.'[35] In total B Company had killed 20 people in the Chuka area on the orders of Griffiths who used authority over the troops to ensure that they carried out unlawful commands.

---

32 David Anderson, 'A Very British Massacre', *History Today, August 2006* (2006), 21.
33 Huw Bennett, *Fighting the Mau: The British Army and Counter-Insurgency in the Kenya Emergency.* (Cambridge: Cambridge University Press, 2012), 189.
34 Ibid., 181.
35 Ibid., 183/4.

Othering

A French anthropologist called Claude Levi-Strauss first argued in 1952 that the savage and civilised mind and human characteristics were the same everywhere regardless of education or culture. When writing for the UN he commented that 'the true contribution of a culture consists, not in the list of inventions which it has personally produced, but in its difference from others' he added that for most people 'other cultures differ from his own in countless ways, even if the ultimate essence of these differences eludes him.'[36] Levi-Strauss sparked the debate about the differences between groups and the term othering came into being. Othering has been defined as 'transforming a difference into otherness so as to create an in-group and an out-group.'[37] The creation of an in-group immediately allows for groupthink as the 'group members of an aggressive in-group are connected to each other in such a way that they are likely to experience identical emotions.'[38] This in turn allow for the dehumanisation of the out-group, this has been noticed to be particularly strong 'when an outgroup is perceived to have dissimilar values to the ingroup' as it is then 'perceived to lack shared humanity and its interests can be disregarded.'[39] Another sign of othering is when the in-group ('us') call the out-group ('them') derogatory names and it appears that they greater the difference the worse the names become.

This differentiation between the in-group and the out-group is very common in the military, which is very quick to assign stereotypes, as this enhances team ethos. The Australian Defence Force doctrine confirms that 'the more 'different' a person is from us, the more negative the attributions we might apply to them.'[40] This is taken further as the more superior a soldier believes they are to enemy soldiers the more serious the crimes. Research has noted that 'soldiers tell themselves that the enemy is an inhuman "kraut," "Jap," "gook," "dink," or "raghead," and, by doing so, hope to remove all natural empathy toward those they aim to kill.'[41] But it is not just the enemy

---

36 Claude, Levi-Strauss, *Race and History* (Paris: UNESCO, 1952), 45.
37 Jean-Fracois, Staszak, 'Other/Otherness', *International Encyclopaedia of Human Geography* (2008), 1.
38 Olusanya, Olaoluwa, *Emotions, Decision-making and Mass Atrocities*. (Farnham: Ashgate, 2014), 104.
39 Haslam, Nick, 'Dehumanization: An Integrative Review', *Personality and Social Psychology Review*, Vol. 10, No. 3 (2006), 255.
40 ADF, Military Ethics, 36.
41 Peter Fromm, Douglas Pryer and Kevin Cutright, 'The Myths We Soldiers Tell Ourselves: and the Harm These Myths Do', *Military Review, September-October 2013* (2013), 60.

that can be put into these categories and abused, it can happen within units. Private Danny Chen was the first soldier of Chinese descent to serve in the 21st Infantry Regiment of the US Army. As the only Chinese American he was singled out by fellow soldiers who an investigation found used racial slurs and insults such as '"gook", "chink", "Jackie Chan", "Soy Sauce", and "dragon lady".'[42] This verbal abuse turned into physical abuse and after 6 weeks of it he committed suicide on 3 October 2011 whilst serving on operations in Afghanistan. As in this case, othering can be the first step in a spiral of violence that can lead to dehumanisation and abuse.

Othering Case Study – My Lai Massacre

The My Lai massacre is a notorious event during the Vietnam War. It involved the killing of between 347 and 504 unarmed South Vietnamese civilians on the 16 March 1968 by units of the US Army's 23rd (American) Infantry Division. The units had been told that the hamlets of My Lai would be full of enemy Viet Cong fighters, but instead contained old men, women and children. The killings started on a small scale, but soon developed into a massacre, with children being killed and women being raped. One of the main reasons why these US Army troops believed that they could do these things was othering. The Vietnamese were seen as an inferior out-group, one soldier explained how 'they weren't our equals. They were, you know, a lower class than us.'[43] They were also known as a series of derogatory names by the US soldiers and the 'use of "gooks", "dinks", "dopes", "slants", "slant-eyes" and "slopes" was not only systemic but endemic.'[44] These accepted names led to what was known as the 'Mere Gook Rule' in which 'it was no crime to kill or torture or maim a Vietnamese because he was a mere gook.'[45] Michael Terry a fire team leader at the My Lai massacre explained how 'a lot of guys didn't feel that they were human beings'.[46] Whilst another solider described how the Vietnamese were 'not people', therefore it didn't 'matter what you do to them' but rather 'they were animals with whom one could not speak or

---

42 Stjepan Mestrovic, *The Postemotional Bully* (SAGE Swifts, 2014).
43 Hugh Thompson, *Moral Courage In Combat: The My Lai Story*. Lecture to Center for the Study of Professional Military Ethics (2003), 15.
44 Tony Raimondo, The My Lai Massacre: A Case Study. School of the Americas, Fort Benning. https://downloads.paperlessarchives.com/p/Eyjx/#:~:text=The%20My%20Lai%20Massacre%3A%20A,the%20school's%20human%20rights%20program.
45 Ibid.
46 Lindsay Drew, 'Something Dark and Bloody', *The Quarterly Journal of Military History*, Vol. 25, No. 1, Autumn (2012), 54.

reason.'[47] This wide spread othering of the Vietnamese people made it easy for US troops to abuse and kill them without feeling that it was unethical or morally wrong.

Dehumanisation

Herbert Kelman first discussed dehumanisation in 1973, he believed that dehumanisation deprived 'both victim and victimizer of identity and community.'[48] The US Army Report into abuse in Iraq stated that: 'dehumanization is the process whereby individuals or groups are viewed as somewhat less than human. Existing cultural and moral standards are often not applied to those who have been dehumanized.'[49] It has more recently been defined as the process of 'stripping people of their dignity and humanity.'[50] The stripping of humanity is done in many ways, giving people numbers rather than names, dressing all people the same and so on. Nick Haslam in *Dehumanization: An Integrative Review* identified that there are two basic ways in which dehumanisation is conducted they are 'denying uniquely human attributes to others represents them as animal-like, and denying human nature to others represents them as objects.'[51] He explained that with animalistic dehumanisation people are given traits of an animal. With this animalistic dehumanisation the emotions of contempt and disgust are a key component. Haslam noted that 'disgust and revulsion feature prominently in images of animalistically dehumanized others: represented as apes with bestial appetites or filthy vermin who contaminate and corrupt.'[52] In simple terms others become '"satanic friends" "degenerates," "vermin," and other bestial creatures. It is easier to brutalize people when they are viewed as low animal forms.'[53] With the second type of dehumanisation, which Haslam calls mechanistic, people are seen as objects. He explains that unlike animalistic dehumanisation, mechanistic 'involves emotional distancing and represents

---

47 Ibid.
48 Herbert Kelman, 'Violence Without Moral Restraint: Reflections on the Dehumanization of Victims and Victimizers', *Journal of Social Issues*, Vol. 23 (1973), 251-261.
49 James Schlesinger, *Final Report of the Independent Panel to Review DoD Detention Operations*. Independent Panel to Review DoD Detention Operations (2004), 5.
50 Hogg and Vaughan, *Social Psychology*, 658.
51 Haslam, 'Dehumanization: An Integrative Review', 252.
52 Ibid., 258.
53 Albert Bandura, 'Moral Disengagement', in *The Encyclopaedia of Peace Psychology, First Edition*, ed. Daniel Christie (New Jersey: Blackwell Publishing, 2012), 4.

the other as cold, robotic, passive, and lacking in depth, it implies indifference rather than disgust.'[54]

A limited amount of dehumanisation is accepted in the military as there is a tendency to call the people that it fights 'the enemy' which hides their humanity. This sanitisation of language to avoid thinking about unethical actions is what Albert Bandura called euphemistic labelling. Using the term of 'the enemy' helps to avoid a soldier having to think about killing other humans. Another euphemism used is 'collateral damage' which is the term used when civilians are accidently killed, which helps to erase the reality of the situation. In its research the ICRC notes that the: 'recourse to euphemisms is commonplace when one refers to war crimes in wartime: people speak of "events", "police actions", "mopping-up operations", "dealing with a target", "surgical strikes" and so on.'[55] All of these allow for soldiers to become emotionally detached from the reality of their actions and the use of detached language eases their feelings about the consequences. In the examination of the unethical case studies, dehumanisation was the one common behaviour that appeared in all cases.

Dehumanisation Case Study – Abu Ghraib Abuse

In 2004 during the early stages of the Iraq War, detainees were housed in Saddam Hussien's old torture jail known as Abu Gharib prison on the outskirts of Baghdad. The 800th Military Police Brigade was responsible for running the prison. The US authorities soon began to use some of the interrogation techniques that had been developed and used at Guantánamo detention centre to try and gain intelligence from the detainees. These techniques included sleep deprivation, nudity, the use of loud music and noise and preying on phobias. Stripping prisoners of their clothes was a common form of sexual humiliation and degradation during the torture at Abu Ghraib. In an analysis of these techniques, it was noted that 'hooding and/or blindfolding the detainees, removing their clothing, and using dogs to elicit fearful confessions from the detainees cumulatively accomplish the dehumanization of detainees.'[56] The study also perceived that 'the public

---

54 Haslam, 'Dehumanization: An Integrative Review', 258.
55 ICRC, *The Roots of Behaviour in War*, 9.
56 Kristin Richardson, *The Social Psychology of Evil: A Look at Abu Ghraib*. Masters Thesis, Clemenson University (2012), 97.

forced removal of clothing depersonalizes, increases actual and sense of vulnerability, humiliates, degrades and dehumanizes detainees. Forcing the detainees to remain nude for extended periods of time teaches and internalizes the belief that they are less than human and undeserving of humane treatment.'[57] It finally concluded that 'in an environment of moral disengagement that countenances authorized techniques designed to humiliate and dehumanize detainees, it is not surprising that other forms of human cruelty such as physical assault were practiced.'[58]

The abuse of detainees was principally committed by 372nd Military Police Company who were responsible for the night shift guard of a group of 50 detainees. The main culprits were Specialist Charles Graner and Private Lynndie England under the command of Staff Sergeant Ivan Frederick. They performed acts which included physical abuse, sexual humiliation, physical and psychological torture and rape. The situation in the prison and the exposure to the interrogation of the detainees had an effect on Graner and England. It was recorded that 'the stripping away of clothing may have had the unintended consequence of dehumanizing detainees in the eyes of those who interacted with them.'[59] This allowed them to start to perform abusive acts on the detainees who they had dehumanised as this 'lowers the moral and cultural barriers that usually preclude the abusive treatment of others.'[60] An example of this was the 'the infamous "leash photograph" … a picture of then Private Lynndie England holding a tank tie-down strap around the neck of a naked Iraqi detainee known as "Gus".'[61] The treating of a detainee as a dog suggests animalistic dehumanisation. In a conclusion to how this abuse happened in Abu Gharib, but not in other prisons in Iraq it was determined that 'if deindividuation and dehumanization are combined (as they were at Abu Ghraib), it results in the degradation, humiliation, torture, and abuse of individuals at the hands of normal, average, mentally stable men and women.'[62]

---

57 Ibid., 98.
58 Ibid., 75.
59 Schlesinger, *Final Report of the Independent Panel to Review DoD Detention Operations*, 7.
60 Ibid.
61 George Mastroianni, 'Looking Back: Understanding Abu Ghraib', *Parameters*, Vol. 43, No. 2 Summer 2013 (2013), 56.
62 Richardson, *The Social Psychology of Evil*, 91.

Demonisation

In 2008 Guy Faure highlighted demonisation in *Negotiating with Terrorists: A Discrete Form of Diplomacy*. In this he stated that 'demonization is the characterization of individuals, groups or political bodies as evil.'[63] The term was well known by 2008 although the concept seems to have first been explored by Harold Lasswell in the 1927 paper *Propaganda Techniques in the World War*. In this Lasswell discussed Satanism, which was the demonisation of the enemy by propaganda, such as the exaggeration of atrocities. In the ICRC's work on demonisation they suggest that 'the humanity of the other side is denied by attributing to the enemy contemptible traits, intentions and behaviours.'[64] This was exemplified when post the 9/11 Terrorist Attacks on the Twin Towers, US President George W. Bush named an 'axis of evil'. This included states such as Iran, Iraq and North Korea who were then demonised by Western governments. Albert Bandura discussed a similar but opposite concept which was termed moral justification. He wrote that people justify violent acts by viewing them as righteous such as 'protecting cherished values, fighting ruthless oppressors, preserving peace, saving humanity from subjugation.'[65] This division of 'good' and 'evil' is a dangerous idea as it can quickly lead to the dehumanisation of the other.

In the volatile, uncertain, complex and ambiguous contemporary battlefield there is an urge to simplify activity so that it can be easier to understand. In these situations, there can be the temptation to believe that 'all we do is good' and 'all they do is bad', intentions and actions become black and white. But combat is never black and white it is always a shade of grey, we do good and bad actions and the enemy do good and bad actions. In this situation 'the enemy is the Other, the complete antithesis of the collective self, the embodiment of absolute evil' in this situation 'the destruction of the enemy has become a moral duty.'[66] This has been reinforced in recent conflict in Iraq and Afghanistan where both Islamic State and the Taliban were demonised by the West. Where this is not recognised there is an impulse to simplify down to stereotypes, which can ultimately lead to dehumanisation.

---

[63] Guy Faure, 'Negotiating with Terrorists: A Discrete form of Diplomacy', *The Hague Journal of Diplomacy*, Vol.3, No. 2 (2008), 193.
[64] ICRC, *The Roots of Behaviour in War*, 10.
[65] Albert Bandura, 'Moral Disengagement', 2.
[66] Anthony Coates, 'Culture, the Enemy and Moral Restraint in War' in *The Ethics of War: Shared Problems in Different Traditions*, ed Richard Sorabji and David Rodin (London: Routledge, 2006), 217.

When this occurs and a group 'dissociates themselves from the humanity of others, they may develop a callous disregard for the rights of others.'[67] In these situations the likelihood of unethical actions against the enemy is greatly increased.

Demonisation Case Study - Voćin Massacre

The Voćin massacre was the killing of Croation civilians by the Serbian White Eagles paramilitary unit. This was during the Croatian War of Independence, in the initial stages of the Balkans War. The White Eagles had occupied the area for four months throughout which 'they treated the civilians in an inhumane way, declining them medications and food.'[68] A court case heard how during the period of 'Serbian occupation Voćin's non-Serb villagers were inhumanely abused and harassed. However, evil incarnate descended upon Voćin on a cold December day.'[69] The massacre was conducted when the White Eagles were ordered to abandon the village on 13 December 1991 before the Croatian Army recaptured the area. On this day up to 43 civilians were killed, all were Croats, except for one 77-year-old Serb who had tried to protect Croat neighbours. The victims were tortured and killed, 'some were hit with hammers, beaten with steel chains, struck with axes or had nails hammered into their chests. Some were also set on fire after they were slaughtered.'[70] When the massacre was over, the White Eagles blew up the 550-year-old Roman Catholic church in Voćin.

When the Serb White Eagles mistreated the local Croats and finally conducted the Voćin massacre, it 'was not a spontaneous event like My Lai, rather it was calculated Serbian policy.'[71] During the wars in the Balkans, Serbs, Croats and Muslim forces all conducted ethnic cleansing operations on populations that they had demonised.

Summary of Behaviours

As explained at the start of chapter 3 there are many behaviours that individuals and teams use every day to function in their daily lives and these

---

67 ADF, *Military Ethics*, 36.
68 Miloš Stanić, '27 Years On, Croatia Tries Serbs for Vocin Massacre', 2018. https://balkaninsight.com/2018/10/15/27-years-on-croatia-tries-serbs-for-vocin-massacre-10-10-2018/.
69 Jerry Blaskovich, 'Odyssey to Dante's Inferno', *Medicinski vjesnik*, Vol. 27, No. 1-2 (1995), 134.
70 Stanić, '27 Years On, Croatia Tries Serbs for Vocin Massacre'.
71 Blaskovich, 'Odyssey to Dante's Inferno', 134.

are not unethical in themselves. However, in the case studies examined there were 12 common behaviours that tended to be conducted by people when they were in a stressful situation and the situational influencers were applied. Each of the common behaviours were explained using an appropriate case study with a view to demonstrating its real-life application. However, in many case studies there were a series of behaviours at work which combined and led to unethical actions. In the next chapter the flow of unethical actions is explained and it will establish how these multiple behaviours coalesce and cause people to act in immoral ways.

5

The Flow of Unethical Actions

Introduction

In 2012, I was the Chief Plans Officer for the UN Military Mission in the Democratic Republic of Congo. The mandate for the mission was the 'protection of the people' and to help achieve this I had established an operation to protect a network of villages near to every UN company base in the contested areas. On a visit to one of these bases, my local Congolese worker told me that the operation was going well, but the company never deployed at night to protect the villagers when they called. I was aware that we had received reports of around 15 patrols a night from that unit, but when I questioned the staff they confirmed that they never left their camp at night. I raised this issue with their brigade commander, the reply was that "I will not lose one of my soldiers for the Congo". I reminded brigade commander of the mandate to protect the people and the fact that the UN troops had armoured vehicles and night viewing devices, whilst the rebels were very lightly armed, however he remained adamant that despite this he would risk no one. On return to Kinshasa, I raised this issue with the force commander, who said that although they should be responding, he would not force this national contingent to uphold the mandate. I was now powerless to enforce the protection of the population in that area. The unit, the brigade commander and the force commander all knew the UN mandate, but they had allowed the power of the situation to stop them doing the right thing and all were now acting in a way that was unethical. Despite their behaviour, I did manage to get South African attack helicopters to answer some of the distress calls from the population in that brigade area to try to off-set the lack of ground patrolling by the units.

The S-CALM Model

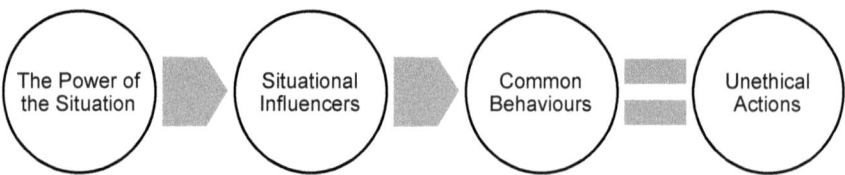

Figure 5.1 Flow of Unethical Actions

Flow of Unethical Actions

When a leader and their team find themselves in a stressful situation, the power of the situation builds up. At the extreme this difficult, stressful situation might be combat, however more routinely it could just be a busy day in the office, when the team need to produce an output under the pressure of time. In trials conducted at RMAS it was discovered that stress was brought on by the requirement to produce an output, by the requirement to do it within a set time and by the requirement to work with a team that was not performing at its best. These all produced a power of the situation and as identified, leaders and teams were then susceptible to situational influencers. In military situations the common situational influencers were identified as a hostile environment, normalisation of violence, weak leadership and a lack of supervision, lack of resource and fatigue and enhanced emotional state. Under the power of the situation, the situational influencers take effect they overlap and interlock, tending to feed off each other. For example, a hostile environment can lead to normalisation of violence and fatigue can lead to enhanced emotions. Once established the situational influencers cause people to use heuristics and bias and behave in unethical ways that they would not normally wish to behave in if they were in a normal, rational state and not in the grip of the situation. These behaviours could lead individuals to conform to a group or act as a bystander when they would normally intervene. It can also lead a team to carry out risky shift and keep pushing the boundary of acceptable behaviour or dehumanise members of an out-group. This chain reaction in stressful situations is termed the 'Flow of Unethical Actions'. Figure 5.1 is a diagram which demonstrates how this flow operates.

This flow tends to be dominated by what Daniel Kahneman called System 1 thinking, which was explained in chapter 3. This intuitive, fast thinking can drive people into actions that they often regret when they have time to reflect on their behaviour. Research has shown that in difficult, stressful situations there is a tendency for cognitive functioning to be restricted as 'stress leads to automatic rather than controlled thinking.'[1] This allows individuals and teams to be driven by their System 1 thinking, bouncing from one step of the flow to the next without considering the consequences of their actions.

The RMAS 2023 research survey asked the question, Is it always best to go with your initial intuitive decision? The rationale for this question was to establish the trainees understanding of the need to balance the use of System 1 and System 2 thinking. Research Question 1 for those that had no previous training, was aimed at the effectiveness of training on the cohort over the year. This cohort remained constant at around a quarter believing that System 1 thinking was favourable, despite their education about the requirement for a balanced decision-making system. Research Question 2, which was for all trainees, 'UK Ocdts, [Officer Cadets] 23% agreed that you should always go with your intuitive decision and of international Ocdts, 20% agreed.' However, the research concluded that 'For both UK and international Ocdts, the majority favoured rational, System 2 type thinking, but a large proportion supported System 1 thinking or were unsure, suggesting some confusion about thinking skills at this stage.'[2] Therefore, even after significant training, there was still some misunderstanding about the effective use of System 2 thinking.

In many cases, without recourse to rationale thinking unethical actions happen without the perpetrators being aware that they are in the spiral of violence.

Militaries have identified this to some extent, but they still tend to align unethical actions as an influence of culture, rather than as a result of the power of the situation. The British Army believes that 'a leader must first understand the culture and climate in which they are operating and how this shapes their own beliefs, values and behaviours, as well as those of the people they are leading.'[3] The Australian Defence Forces maintains that 'unethical or otherwise unacceptable behaviour in a team always represents a disconnect between our Defence values and the team's culture.'[4] The US Army only comments on the effect after the act stating that 'unethical behaviour quickly destroys organizational morale and cohesion—it undermines the trust and

---

1 A. J. Porcelli and M. R. Delgado, 'Acute Stress Modulates Risk Taking in Financial Decision Making', *Psychological Science*, Vol. 20, No, 3 (2009), 283.
2 Dennis Vincent and Alex Muhl-Richardson. *An Analysis of the Effectiveness of the S-CALM Model of Ethical Leadership at the Royal Military Academy Sandhurst. Sandhurst Occasional Paper No. 37.* (Sandhurst: Central Library, 2024), 18.
3 MOD, *Army Leadership*, 1-8.
4 ADF, *ADF Leadership*, 28.

confidence essential to teamwork and mission accomplishment.'[5] Although these statements about ethos and values are generally correct, they don't explain how the disconnect between culture and behaviour takes place or how it affects teams actions.

A case study that underlines how the 'Flow of Unethical Action' can cause well trained soldiers to commit acts that they would later regret is that of Sergeant Alexandera Blackman. Sergeant Blackman was a Royal Marine serving in Afghanistan in 2011 who shot and killed an injured Taliban fighter. The case study will firstly set out the background to the shooting, then explain how the situational influencers built up and caused changes in Sergeant Blackman's behaviour, thinking and decision making. Finally, the common behaviours that took effect will be explored and outline why the shooting wasn't stopped by other members of the patrol.

Flow of Unethical Actions Case Study: Sergeant Blackman Killing

Sergeant Blackman was a member of J Company, 42 Commando, Royal Marines deployed in Afghanistan on Operation Herrick 14 in 2011. He was in command at Check Point Omar, Nad-e Ali District, Helmand Province. On 15 September 2011 Sergeant Blackman had been leading a patrol when it was re-tasked to investigate a Taliban insurgent wounded by an Apache helicopter. The patrol initially followed the correct action and gave first aid but after some discussion Sergeant Blackman ordered the Afghan to be moved out of sight of the surveillance system, a camera on a balloon above British Forward Operating Base Shazad and the Apache helicopter which was still overhead. Once out of sight of these observers first aid treatment was stopped. After some discussion the helmet camera video, which recorded the whole incident, shows how Sergeant Blackman,

> crouches down and aims his pistol at the centre of the insurgent's chest, fires once at point blank range, and immediately stands back up. The insurgent's legs, which are bent at the knee, begin to move left and right. His upper body and arms then start to writhe, and his

---

[5] US Army, *ADP 6-22 Army Leadership And The Profession*, 2-6.

head shakes back and forth. His breathing starts to become laboured.⁶

Sergeant Blackman killed the Afghan at this point and the patrol made their way back to Check Point Omar. However, the killing had been filmed on the helmet camera and later this video was found by the police and a court martial and subsequent Appeal Court hearing were conducted.

Situational Influencers

The main material used to examine the situational influencers in this case study came from three main sources: the transcript of the helmet camera, evidence given at the trial and appeal of Sergeant Blackman and the Royal Navy's Telemeter Report which was an internal review into the Sergeant Blackman incident. The Telemeter Report noted that 'current training appears to be lacking in terms of understanding how the situation demands can undermine rules and regulations, particularly in situations where "the battle lines are not straight".'⁷ In addition to this comment, the Report outlined some of the five common situational influencers, all of which will now be explored to identify how they manipulated Sergeant Blackman to shoot the Afghan.

Hostile Environment. The Nad-e Ali District had a strong insurgent presence and attacks on the Royal Marines were frequent. Every patrol could expect the enemy 'to provoke a kinetic response.'⁸ Sergeant Blackman recalled the stress and how 'the day in, day out threat that every step you take on patrol might be your last' affected the Marines and how this increased with activities like 'stones thrown by children that could suddenly be grenades.'⁹ During the Court of Appeal evidence, Marines from Sergeant Blackman's patrol explained how they were 'under constant external threat' and how they were 'always on edge and did not feel safe at night.'¹⁰ The hostile environment was explained in more detail at Court of Appeal, it was noted that 'it was difficult to detect insurgents in the local population whose hearts and minds they were seeking to win over.' It was also explained that 'ambushes by insurgents and the threat of IEDs was constant; statistics

---

6 Royal Courts of Justice. Court of Appeal Judgement in the case between Regina and Alexander Wayne Blackman, 7 and 8 February 2017, Para 22.
7 Navy Command, Telemeter-Internal Review, Portsmouth: Navy Command Headquarters, 2014, C1.
8 Navy Command, Telemeter-Internal Review, B1.
9 Alexander Blackman, *Marine A: My Toughest Battle.* (London: Mirror Books, 2019), 175.
10 Royal Courts of Justice, Court of Appeal Judgement, Para 99.

showed that there was an IED explosion (whether controlled or instigated by the insurgents) on average every 16 hours.'[11] Sergeant Blackman reflected that when they were sent to investigate the wounded Taliban, he was concerned about treating the insurgent and calling in a medivac helicopter as this might subject the patrol to a follow up attack. He said that 'the longer we were out here, the greater the risk of being exposed. The lads did not deserve that.'[12] Therefore the hostile environment was influencing Sergeant Blackman's thinking and decision making at this critical time and causing behaviour he would later regret.

Normalised Violence. The Telemeter Report noted that 42 Commando had suffered a high rate of attrition during their deployment. The unit had eight members killed in action: Lieutenant Ollie Augustine, Sergeant Barry Weston, Lance Corporal Martin Gill and Marines Sam Alexander, Nigel Mead, James Wright and David Fairbrother.[13] In fact, 'Blackman himself had almost been killed in a grenade attack a month before the incident, and had developed a deep paranoia about his own safety.'[14] Nevertheless, the death of Lieutenant Augustine really hit Sergeant Blackman hard as this was his troop commander and once he was killed Blackman had to take over the responsibilities of commanding both the troop and Check Point Omar. In response to this attrition the Marines had normalised violence and had increased their aggression. The new commanding officer of 42 Commando, Lieutenant Colonel Oliver Lee noted how J Company was 'employing a "very high level of violence" through the heavy use of mortar bombs when minimum force was meant to be the order of the day. He claimed a senior officer who visited J Company described it as "psychologically defeated, bereft of ideas, unpredictable and dangerous".'[15] This description could have been used for Sergeant Blackman at this time, with the troop commander killed and a recent near death experience he recalled how it was 'the deaths of young men like Lieutenant Augustin [sic] and Marine Sam Alexander that dominated the mind.'[16] The high level of hostility had created a

---

11 Ibid.
12 Blackman, *Marine A*, 169.
13 'Royal Marines from 42 Commando return from Afghanistan', BBC News,, 25 October 2011, https://www.bbc.co.uk/news/uk-england-devon-15448969.
14 Tom McDermott, 'We Need to Talk About Marine A: Constant War, Diminished Responsibility and the Case of Alexander Blackman', Canberra: ACSACS Occasional Paper No. 6, 2017, 2.
15 'Alexander Blackman's company was out of control, claims former comrade', The Guardian, 15 March 2017, https://www.theguardian.com/uk-news/2017/mar/15/alexander-blackmans-company-was-out-of-control-claims-former-comrade.
16 Blackman, *Marine A*, 125.

normalisation of violence that affected Sergeant Blackman's decision making and behaviours.

Lack of Supervision. The Telemeter Report noted that Check Point Omar was geographically isolated and there was a lack of oversight by the chain of command which allowed Sergeant Blackman to break Standard Operating Procedures (SOPs). The Report recommended that 'good leadership, effective oversight and firm supervision are enduring requirements in every leader', but it considered that 'a number of those involved in this incident both directly and indirectly, felt that the chain of command had failed to provide them with adequate support.'[17] The Brigade Commander had some concerns about the leadership of 42 Commando by its commander Lieutenant Colonel Ewen Murchison. In addition, Lieutenant Colonel Lee, who at this time was commanding 45 Commando raised 'concerns over 42 Cdo's approach' but the Brigade Commander 'did not formally reproach CO 42 over his units' culture.'[18] Lieutenant Colonel Lee recalled,

> The leadership and oversight of Sgt Blackman by his commanders Lt Col Murchison and Maj [Aaron] Fisher was shockingly bad, and directly causal to Sgt Blackman's conduct … Murchison, who was Blackman's commanding officer for the vast majority of the six-month tour, only visited the sergeant's command post once or twice, leaving him feeling isolated.[19]

The Court of Appeal's conclusion recorded this lack of supervision had 'intensified the feeling of isolation at CP Omar.'[20] Due to the isolation and command culture, the unit's SOPs were seen as optional. The Telemeter Report stated that 'it was evident in this case that the situational demands had undermined clear Standing Operating Procedures.'[21] When Lieutenant Colonel Lee took command of 42 Commando, he visited Check Point Omar and stated that 'on entering the base, I found that it had a feral and squalid air about it. The men in the base were disrespectful, surly, dressed incorrectly and there was litter and rubbish everywhere.'[22] The lack of supervision,

---

17 Navy Command, Telemeter-Internal Review, B1. Navy Command, Telemeter-Internal Review, B2.
18 Ibid., B1.
19 *The Guardian*, 'Alexander Blackman's company was out of control, claims former comrade'.
20 Royal Courts of Justice, Court of Appeal Judgement, Para 101.
21 Navy Command, Telemeter-Internal Review, B1.
22 *The Guardian*, 'Alexander Blackman's company was out of control, claims former comrade'.

increased by the death of Lieutenant Augustine, had 'allowed professional standards to slip to an unacceptable level in CP [Check Point] OMAR.'[23] This lack of supervision in turn resulted in a tendency for ethical drift to set in which led to the ignoring SOPs and a relaxation of standards of discipline and behaviours, such as the correct treating of wounded combatants.

Weak Leadership. In addition to a lack of supervision there was also weak leadership. The Telemeter Report noted the inappropriate command supervision and it has also been noted that 'Blackman's decline was reflected in worsening discipline and conditions at CP Omar.'[24] The Court of Appeal stated that 'the perceived lack of leadership by Col Murchison and Major Fisher and their perceived lack of support were key stressors.'[25] Although Sergeant Blackman was an experienced senior NCO, he was not trained to act as the troop commander and needed support and guidance from the chain of command which he did not receive. The Telemeter Report noted that in this type of situation 'strong leadership and regular oversight is required.'[26] However, Sergeant Blackman did not set a good example as the patrol commander and the report stated that 'his poor leadership was a significant contributory factor in the way the insurgent was treated by other members of the patrol.'[27] Sergeant Blackman had influence over the patrol members, and when Blackman's own behaviour fell short of the acceptable standard this was reflected in the behaviour of those he commanded as the leader.

Lack of Resource. The common lack of people and time were evident in this case study. The Appeal Court identified that 'the multiple at CP Omar was undermanned' it went on to describe how 'the previous multiple had been 25' but the multiple under Sergeant Blackman 'was 16.'[28] Other analysis established that 'the manning of CP Omar was reduced from 25 to 16 for 42 Cdo, which increased patrolling tempo and isolation.'[29] The escalation in patrolling tempo meant that 'the multiple was required to patrol between 5 and 10 hours a day over rough ground in heat that was normally over 50 degrees Celsius when carrying a minimum of 100lbs

---

23 Navy Command, Telemeter-Internal Review, B1.
24 McDermott, 'We Need to Talk About Marine A: Constant War, Diminished Responsibility and the Case of Alexander Blackman', 4.
25 Royal Courts of Justice, Court of Appeal Judgement, Para 102.
26 Navy Command, Telemeter-Internal Review, C1.
27 Ibid., B1.
28 Royal Courts of Justice, Court of Appeal Judgement, Para 99.
29 McDermott, 'We Need to Talk About Marine A: Constant War, Diminished Responsibility and the Case of Alexander Blackman', 4.

of equipment.'[30] Therefore as a 'consequence the men became physically very tired, particularly at times of illness or insurgent activity.'[31] With the added burden of command 'Blackman was significantly deprived of sleep.'[32] There was an additional time pressure during the incident which added to the power of the situation. Blackman was concerned that remaining static at the scene of the shooting would put the patrol in danger. There had already been considerable 'radio chatter [that] suggested another attack was imminent; the second insurgent had not been accounted for; there was therefore a heightened threat of attack.'[33] Therefore in Blackman's fatigued mind the ability to make the correct moral decision was confused with the perceived tactical need to extract from what he believed to be a dangerous situation.

Enhanced Emotion State. When Sergeant Blackman found that the Afghan was still alive, he was heard to say 'why couldn't (you) just be fucking dead.'[34] He later admitted that he was both 'frustrated' and 'angry' that the insurgent was not dead and that he would now have to deal with the Afghan.[35] Sergeant Blackman's attitude was described as being 'angry and vengeful and [he] had a considerable degree of hatred for the wounded insurgent.'[36] This attitude influenced the other patrol members who took out their frustration and anger on the Afghan. On the video 'one of the Marines leans towards him and says "you're browners fella" [slang: browners = brown bread = dead]' a little later the Marine 'says "don't give a fuck about you son".'[37] Just before he shot the insurgent Sergeant Blackman recalled that 'frustration was boiling over' and that 'I was taking out my frustration on this mortally wounded insurgent.'[38] Emotions were running high during this incident, the patrol was fatigued, scared, frustrated and angry and the consequences of these emotions were taken out on the wounded Afghan.

---

30 Royal Courts of Justice, Court of Appeal Judgement, Para 99.
31 Ibid.
32 McDermott, 'We Need to Talk About Marine A: Constant War, Diminished Responsibility and the Case of Alexander Blackman', 2.
33 Royal Courts of Justice, Court of Appeal Judgement, Para 104.
34 Ibid., Para 21.
35 Blackman, *Marine A*, 172.
36 Royal Courts of Justice, Court of Appeal Judgement, Para 102.
37 Ibid., Para 21.
38 Blackman, *Marine A*, 176. Blackman, *Marine A*, 179.

Common Behaviours

The situational influencers created the environment where Sergeant Blackman and the patrol were susceptible to behaviours which led to unethical actions. During this incident many common behaviours explained in earlier chapters were displayed. However, the dominant ones that were the most conspicuous were authority bias, dehumanisation, cognitive dissonance, risky shift and bystander intervention.

Authority Bias. With the troop commander killed, Sergeant Blackman was in sole command of the troops at Check Point Omar and as already identified there was a feeling of isolation from the remainder of J Company that would have enhanced Blackman's leadership. The Court of Appeal noted that the 'perception of the lack of leadership from those above heightened his sense of Responsibility.'[39] The patrol members were obedient to Sergeant Blackman's authority as they identified that he was trying to look after them despite a general feeling that they were not being cared for by the system. The Telemeter Report identified this and stated that 'Blackman's rank was a significant contributory factor in preventing others within the patrol from questioning his orders.'[40] This in turn generated loyalty by the troops to Sergeant Blackman and increased the authority bias that was already in place. Therefore, when he gave directions about the wounded Taliban, for example to move the Afghan out of sight of the surveillance systems and to watch out for the Apache helicopter, these orders were followed without question.

Dehumanisation. By the stage of the deployment when the patrol came across the injured Afghan, they had dehumanised the Taliban fighters. Lieutenant Colonel Lee stated that 'some marines operating in Helmand in 2011 were guilty of dehumanising the enemy, a state of mind that in the past had led to atrocities such as the My Lai massacre in Vietnam.'[41] From the very start of the video recording the Afghan is treated badly, he is dragged, kicked and thrown to the floor as he is moved to the side of the field. Once in cover, Sergeant Blackman noted that 'one of the lads began to go through the medical processes in which we were all trained, dressing what wounds we could, attempting to stabilise him.'[42] However, while doing this action the Marine can be heard to say '"For fuck's sake, I cannot believe I'm doing

---

39 Royal Courts of Justice, Court of Appeal Judgement, Para 102.
40 CAL, What leaders are, B2.
41 *The Guardian*, 'Alexander Blackman's company was out of control, claims former comrade'.
42 Blackman, *Marine A*, 173.

this". Another Marine replies, apparently referring to the helicopter, "Wait a minute, just pretend to do it, till he's [the helicopter] behind them trees".'[43] The handling and shooting of the Afghan ultimately demonstrate a disregard for their humanity, which was brought on by the power of the situation and the subsequent situational influencers mentioned early. The Court of Appeal summarised Sergeant Blackman's dehumanisation and the 'consequence was that he had developed a hatred for the Taliban and a desire for revenge.'[44]

Cognitive Dissonance. There was a friction in Sergeant Blackman's mind about what to do with the wounded Afghan. He knew that they should have given first aid to the insurgent and then called in a medical evacuation helicopter to take the Afghan back to base. Nevertheless, the easy option was to wait until the Afghan was dead and then to return to Check Point Omar as quickly as possible before any further attacks took place. The Telemeter Report noted that Sergeant Blackman had difficulty in 'changing from a mind-set which required him to kill an enemy to one which accepted having to administer first aid to an enemy in order to save his life.'[45] As the time slipped away Sergeant Blackman recalled 'how desperately we wanted to be back in our compound.'[46] It was at this point that Sergeant Blackman shot the Afghan and the video transcript noted that 'after watching for 15 seconds and adjusting his backpack, the appellant [Blackman] says "There you are, shuffle off this mortal coil, you cunt." After another 10 seconds he says to the writhing, dying Afghan "It's nothing you wouldn't do to us." He then addresses the patrol and says "obviously this doesn't go anywhere, fellas. I've just broke the Geneva Convention."'[47] Sergeant Blackman knew the correct actions to take but made the decision to shoot the Afghan despite identifying that he understood the rules of engagement and international humanitarian law, he still breached them under the power of the situation.

Risky Shift. The patrol demonstrated groupthink in all wanting to get back to base, but this developed into risky shift. As has been identified the spiral of violence began with Sergeant Blackman's frustration and anger that the Afghan was still alive. The next step was the rough handling of the Afghan as he was moved to the side of the field. Sergeant Blackman said at this point "get him closer in so PGSS [surveillance camera] can't see what we're doing to him." At this stage the 'insurgent is moved a few more feet

---

43 Royal Courts of Justice, Court of Appeal Judgement, Para 22.
44 Ibid., Para 109.
45 *The Guardian*, Alexander Blackman's company was out of control, claims former comrade'.
46 Blackman, *Marine A*, 176.
47 Royal Courts of Justice, Court of Appeal Judgement, Para 22.

and, after some discussion, a marine drags him again and throws him down on his back in a clear area at the side of the field, close to the tree line. The insurgent winces in pain.'[48] Once under cover there was a discussion about giving first aid which developed as follows,

> Blackman says, "Anybody want to do first aid on this idiot?", to which others respond "No". The guard continues to point his pistol at the insurgent. Somebody says, "I'll put one in his head if you want" and someone else makes another suggestion (the detail of which is inaudible). There is then laughter, and someone says, "Take your pick". After a few seconds the appellant [Blackman] comes closer and stands over the body, saying "No, not in his head, 'cause that'll be fucking obvious".[49]

Just after this exchange about shooting the Afghan 'a marine says "Al, just strangle him" to general laughter.'[50] The patrol quickly moved from legitimate action to the shooting and this downward spiral can be tracked in the comments on the video as under the power of the situation they encourage each other to consider the next unethical step.

Bystander Effect. The Telemeter Report raised the concept of the bystander effect and the idea of misplaced loyalty to a group when it stated that 'loyalty to an "oppo" is best expressed by challenging him before he makes a mistake rather than trying to cover-up for him afterwards.'[51] It went on to make a recommendation for the future and advised that patrols should have 'at least one individual with the necessary prestige and confidence to challenge the commander, when appropriate.'[52] In the Sergeant Blackman case the individual that had the prestige and confidence to do this was most likely Corporal Christopher Watson, one of Sergeant Blackman's NCOs, but he was also a co-defendant in the courts martial. Corporal Watson was acquitted of any crime and said that the comments he made which encouraged Blackman were just bravado. However, Watson's support for the treatment of the Afghan and the encouragement he offered towards mistreatment meant that no other member of the patrol would have felt empowered to speak up against the actions. The Telemeter Report noted that 'group conformity

---

48 Ibid., Para 21.
49 Ibid.
50 Ibid., Para 22.
51 Navy Command, Telemeter-Internal Review, C1.
52 Ibid., C3.

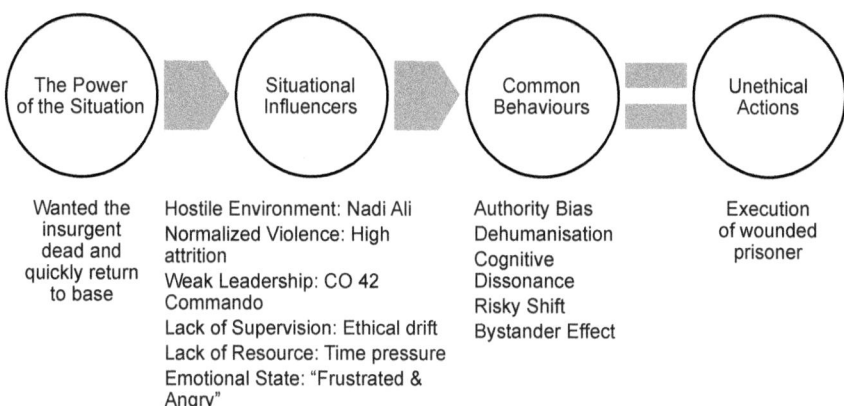

Figure 5.2 The Flow of Unethical Actions – Sergeant Blackman Case Study

and positive past experiences of Sgt Blackman may also have contributed' to a lack of intervention.[53] With Sergeant Blackman and Corporal Watson approving the treatment of the Afghan the diffusion of responsibility within the patrol restricted any of the marines from intervening to stop the spiral of violence from taking place.

Summary

The Sergeant Blackman case study is a useful one because it is contemporary and has had an official military review and a series of court martials. It demonstrates how the flow of unethical actions can take place in a stressful situation. Figure 5.2 lays out how the flow happened in this case.

The case study is also excellent at demonstrating how System 1 thinking can drive the actions of a leader and their team when the power of the situation has an effect on them. Sergeant Blackman recalled that 'some things have happened before you really understand them at all.'[54] He was under pressure and unthinkingly shot the Afghan. He recalled that immediately after 'I knew I'd done something stupid almost straght away.'[55] He also recollected how,

> You make a split second decision on the day, there's the gunshot, there is a 15-20 second pause before I say

---

53 Ibid., B2.
54 Blackman, *Marine A*, 178.
55 Ibid., 179.

anything after that, when your mind is racing you have done something stupid. Why have you done what you have just done?[56]

It is at this stage, during that pause that Sergeant Blackman's System 2 thinking is triggered and he recalled that 'there is the pause when I went through that realisation that I had made a mistake.'[57] He reflected how he tried to 'justify what you've done and you seek reassurance from those around you. Cause you know you've done something wrong and this is going to go badly why have I done it, you're not going to drop me in it are you lads.'[58] Later when watching the video in a police station he reflected how 'it had seemed so brutal to me.'[59] He finally concluded that it was 'not something I am proud of, I let myself down.'[60]

In attempting to understand the incident a senior Royal Marine officer Brigadier Bill Dunham, emphasised how the shooting was 'not consistent with the ethos, values and standards of the Royal Marines.'[61] The Telemeter Report also concluded that the Values and Standards 'define our ethical approach and underpin our legitimacy' but continued that 'this case illustrates, the comtemporary and future operating environments are likely to pose highly complex problems'. It finishes that a deep understanding of the Values and Standards are required in the future if they are to 'apply their principles under great stress on operations.'[62] Nonetheless, as has been identified, the Values and Standards alone do not give a leader and their team the toolkit required to prevent the power of the situation from affecting behaviours. The next chapters will explain how the S-CALM model can assist commanders in these difficult, stressful situations.

---

56 'Alexander Blackman: How should crimes on the battlefield be handled?', BBC Hardtalk, BBC World Service, 22 January 2020. https://www.bbc.co.uk/sounds/play/w3csy9fj.
57 Ibid.
58 Ibid.
59 Blackman, *Marine A*, 174.
60 BBC, 'Alexander Blackman: How should crimes on the battlefield be handled?'.
61 'Royal Marine Guilty Of Murdering Afghan Fighter', Sky News, Friday 8 November 2013, https://news.sky.com/story/royal-marine-guilty-of-murdering-afghan-fighter-10428900
62 Navy Command, Telemeter-Internal Review, C3.

6

S-CALM:
S – Situational Influencers

Introduction

In 1998, I was commanding a company in West Belfast Northern Ireland. It was a busy operational tour and I knew that I was responsible for the lives of 120 young soldiers and that I would be required to make important decisions on the spot that could result in the death of these young soldiers. This requirement for immediate decision-making weighed on my mind and I decided that I would need to sleep eight hours a night to always be fresh. Of course, I did not sleep like this every night, I went on night patrols or checked sentries, but on most nights, I would hand over to my second in command at midnight and he would wake me if he needed me otherwise I would sleep until 0800 hours when I would rise for breakfast. This took discipline, as there is always that nagging doubt that you should be seen or engaging with others. Nevertheless, this strategy worked well and when required to plan or make decisions I was always in a cognitively good state. In later conflicts, in Afghanistan for example, I was unable to sleep enough, but I had an understanding that my decision-making was not as effective as it should have been and made allowances for this.

Before a leader even considers the S-CALM Model, they must understand that they will be placed in stressful situations and the effect of this will be the application of external pressures that exacerbate their cognitive behaviours. Once they have accepted that their behaviour will be affected, they can start to consider how to stop this happening. The first element of applying the S-CALM model is to recognise and mitigate the situational influencers. The key to this is to understand the influencers that will exacerbate behaviours, recognise them and mitigate their effects. Recognising these influencers and how to mitigate them is a fundamental skill for an ethical leader and military doctrines comprehend the need for it. British Army doctrine states that 'for leaders to be able to lead effectively, they must understand the environment they are operating within as this will influence the decisions they make and shape the way they lead.'[1] The US Army also believe that 'the most effective leaders adapt their approach to the mission, the organization, and the situation.'[2] It adds that 'the situation affects which actions leaders take.'[3] Finally the Australian Defence Force asserts that 'leaders at all levels also have a responsibility to mentor and educate their personnel on the ethics of a particular conflict or situation.'[4]

Recognise Situational Influencers - before the Situation

Some influencers can be recognised before entering the stressful situation if leaders develop the capability to horizon scan for problems. Reflection is important as if a leader 'enters combat without taking time to consider where they stand morally and ethically, it is already too late.'[5] For example, with reflection a leader might know that they will be entering a hostile environment, not have the right levels of supervision or that they will not have enough resource such as time or people. If these influencers can be recognised, some mitigation of them can take place while still in this rational thinking space. If a leader believes that they are going to enter a hostile environment, they can consider how this will manifest itself and what effect it might have on them and their troops. They can then contingency plan to try and avoid an escalation of violence and how they might conduct de-escalation once the violence has ended. If on the other hand they anticipate that there might be a lack of supervision they can ensure that they set a personal ethical example, layout a clear vision and engage with soldiers to confirm that they understand it. It is also important to ensure that subordinates are then empowered to work within the leader's ethical intent. A leader might anticipate that there will be a lack of resource and fatigue, the UK Ministry of Defence identifies that if there is not adequate resource when deployed 'the operation will fall

---

1 MOD, *Army Leadership Doctrine*, 3-6.
2 US Army, *ADP 6-22*, 1-8.
3 Ibid., 1-10.
4 ADF, *Military Ethics*, 5.
5 Mitchell Hall, 'Why Leaders need a Morality Check', *United States Navel Institute*, Vol. 132, No. 4 (2006), 68.

short or, fail.'[6] Conducting an estimate of the operation and planning is vital in order to reduce the physical stress on soldiers once deployed. An estimate will ensure that the task is properly logistically supplied and that there is sufficient workforce to complete the mission. If these are not available, at least the leader will understand that the shortfall will be a friction point. Finally, a leader must understand that sleep is vital to all for both cognitive and physical robustness. As seen in chapter 2, people need to get eight hours sleep a night to retain cognitive function, which reduces in ratio with the number of hours less sleep people have. Therefore, soldiers need to be as rested as possible before operations in order to be able to make ethical decisions in the heat of the moment.

Recognise Situational Influencers - in the Situation

Other influencers may only be anticipated or noticed in the stressful situation. All five of the influencers could be at play in these situations and leaders need to be able to recognise them. Later chapters will encompass the scope of mitigating these influencers, but some elementary concepts are covered here for each of the common five influencers. To highlight how these situational influencers can be recognised and mitigated a single operation has been selected, in this case the British Army in Gorazde in 1995. The United Nations in Bosnia had declared three eastern safe havens for Bosnian Muslim's in Srebrenica, Zepa and Gorazde. The British Army were tasked to protect some 45,000 Bosniaks in the town of Gorazde and the mission was given to the 1st Battalion, the Royal Welch Fusiliers. They had three companies and B Company was commanded by Major Richard Westley. On the 28 May 1995, B Company managed to stop the Serbs from taking the town and prevented a massacre. Major Westley had to act quickly and decisively to fight off the Bosnian Serbs as he knew that the fate of population depended on it. This was later confirmed when the Bosnian Serbs massacred 8,000 Bosniak men and boys in Srebrenica.

Hostile Environment

A hostile environment is one in which a person feels under threat or uncomfortable due to a perception of danger and as mentioned earlier this

---

[6] MOD, *The Good Operation: A handbook for those involved in operational policy and its implementation.* (London: HMSO, 2017), 41.

in some circumstances might be predicted. In many situations this adverse environment may not be forecast and it will need to be recognised. Some hostile environments will be obvious, such as when in combat, however in other less violent situations distinguishing it might be less simple. In this case, individuals should identify the physiological changes such as increased blood pressure, heightened heart and respiratory rate, perspiration and a tightening of the throat. One of the physiological changes that can happened to individuals is the heightening of the ability to focus and this can be used to assist in the mitigation of a hostile environment.

A way to mitigate the physiological changes that can lead to unethical behaviours is to control their effects by using breathing. The US Navy Seals use a breathing technique called Combat Tactical Breathing, which involves taking three to five breaths. For each one they use the following procedure: 'Breathe in counting 1, 2, 3, 4. Stop and hold your breath counting 1, 2, 3, 4. Exhale counting 1, 2, 3, 4.'[7] This breathing exercise allows a person to counter the commonplace physiological changes and take control of their cognitive processing permitting System 2 thinking to start being used.

In the 2023 RMAS research survey a question on the hostile environment was asked, it read, A hostile environment will make you behave in a different way than a safe environment? The rationale for this question was to establish if the officer trainee understands that a hostile environment influences peoples' behaviours. Research Question 1, for those that had no previous training, was aimed at the effectiveness of training on the cohort over the year. This cohort demonstrated a high understanding of the effect of a hostile environment with 98 per cent understanding its influence. There was a similar result in the answers to Research Question 2. In this 'Responses from all Ocdts [Officer Cadets] at the final timepoint showed broad agreement that a hostile environment will influence behaviour, with 99% of UK Ocdts and 93% of international Ocdts agreeing.'[8] These responses demonstrated an excellent understanding of the influence of a hostile environment on behaviour.

---

7 'Combat Tactical Breathing', US Navy, https://www.med.navy.mil/Portals/62/Documents/NMFA/NMCPHC/root/Documents/health-promotion-wellness/psychological-emotional-wellbeing/Combat-Tactical-Breathing.pdf.
8 Vincent and Muhl-Richardson, *An Analysis of the Effectiveness of the S-CALM Model of Ethical Leadership*, 14.

Hostile Environment Gorazde Case Study

Gorazde had two main factors that made it a hostile environment for the soldiers of the Royal Welch Fusiliers, the first was the aggression from the Bosnian Serbs and the second was the ever-present danger of land mines.

There was a lot of Bosnian Serb Army activity aimed against Gorazde. For example, between '17th – 24th April – 400 firing incidents and 50 heavy weapon violations' took place.[9] As May went on the aggression levels raised and on '25th May – Serbs threaten to shell Gorazde camp.'[10] Major Westley recalled a meeting with the Bosnian Serb Army liaison officer, Captain Kepic, at lunchtime on the 28 May. At this meeting 'Kepic stated that if he received orders to take any OPs, the UN soldiers would be well treated. Major Westley replied that any such attempt would be resisted.'[11] Just after this the Bosnian Serbs attacked and the Commanding Officer, Lieutenant Colonel Jonathon Riley, ordered Major Westley to 'defend himself if attacked.'[12] Major Westley remembered that the '28 May was fairly desperate, there were about 45 of us with limited ammunition, but no limited will to fight.'[13] When the fighting was over, Major Westley explained how he de-escalated the company following the fighting, saying that 'we had always been quite good at de-escalating our engagements with the Serbs. De-escalation had become routine, we would be shot at, hit them with everything we had and then negotiate.'[14] Major Westley had built up resilience before the day of combat came and had accustomed the troops to rapid de-escalation of violence.

The Gorazde area was covered with mines, they were mainly made of wood so they could not be located and they fired a bullet up in to the body. Some areas were known to be mined, but other locations were unknown. It created a sense of dread whenever the Royal Welch Fusiliers had to patrol unknown areas. Then on 19 April 1995 B Company had their first mine strike incident, when a 'a four-man foot patrol was moving through the Total Exclusion Zone just outside Gorazde' and 'entered a minefield.'[15] Fusilier Thompson 17 had trodden on one of the mines and got a bullet in the face.

---

9 Jonathon Riley, *White Dragon: The Royal Welch Fusiliers in Bosnia*. (Wrexham: The Royal Welch Fusiliers, 19950, 110.
10 Ibid.
11 Ibid., 50.
12 Ibid., 46.
13 Richard Westley, Interview with author, 2 October 2023.
14 Ibid.
15 Riley, *White Dragon*, 31.

The remainder of the patrol thought he had been shot and took cover. As they did so 'Corporal Jones 10 had trodden on another mine and its bullet had gone straight through his foot' and Fusilier Mee had a bullet that had 'not only gone through his foot but continued up into his chest.'[16] Following this Cpl Williams 49 cleared a path through the minefield and extracted the wounded. Major Westley recalled that the idea of stepping on a mine was 'really an unnerving feeling; one you have to force out of your mind … and yet it's impossible to force the fear out of your mind entirely.'[17] He explained that he dealt with this by ensuring that for any 'patrols going out to areas we believed there were minefields, I would go out with them as morally I couldn't ask them to do it unless I did it also.'[18] By showing that he could cope with the fear and being an exemplar, he was demonstrating to the soldiers that he had mastered the hostile environment.

Normalised Violence

When people are exposed to regular violence it can be seen as an immutable part of daily life and can soon be accepted as the social norm. It can be recognised when there is the general acceptance of violence as being routine and inevitable. This in turn can be manifest by a level of callousness and a lack of compassion to others. Normalised violence can also be identified as a key component of ethical drift when there is no correction of the lack of compassion. The language used at this time can be a way of ascertaining a slip in acceptable behaviour and violence has become the social norm. Some examples, but not an exhaustive list, of this language might be:

- Everyone is doing it
- No one cares
- No one is looking
- We don't have time to do it that way
- I know what we should do but …
- I think this is OK?
- It will make it easier if …
- This is only a small thing.

---

16 Richard Westley and Mark Ryan. *Operation Insanity: The Dramatic True Story of the Mission that Saved 10,000 Lives.* (London: John Blake Publishing, 2016), 39.
17 Ibid., 49.
18 Westley, Interview, 2 October 2023.

A key element in countering the normalisation of violence is for the leader to ensure they retain respect for others, this includes their own followers as well as any adversaries or the local population. This maintaining of respect counteracts the creep of callousness. As an operational deployment goes on a leader can also impede ethical drift by resetting the units' social norm ensuring that respect for others is at the forefront of soldiers' minds. Finally, after a time of trauma, such as in a major combat scenario, there must be time set aside for counselling in order to allow people to process their experiences. These mitigations will assist in limiting subsequent unethical behaviours.

Normalised Violence Gorazde Case Study

There was a great deal of violence during the Bosnian War especially with inter-sectarian aggression. The Bosnian Serbs had begun to kill Bosniak boys, who they reasoned would grow up to become fighters. Major Westley recalled an incident when they shot a ten year old boy named Adem whilst he was on patrol. He said 'I was an experienced soldier; this was the most cold-blooded piece of nastiness I'd seen ... This was deliberate; a murder, right in front of a British UN peacekeeper.'[19] He went to complain to Captain Kepic, the Serb Army liaison officer stating that 'this child was cycling home from school when he was murdered by a Bosnian-Serb solider' Kepic replied 'What is it that you wanted to see me about that was so important?'[20] However, Major Westley recalled how 'they'd picked the wrong peacekeeper to taunt with their casual savagery.'[21] It was from this incident that Major Westley used the tactic of responding to casual acts of Bosnian Serb violence with overwhelming force and then rapid de-escalation. Major Westley described that keeping violence under control began with 'clear orders, and rehearsing rules of engagement. I told them I wanted to absolutely maximise violence when dealing with the threat and then break contact.'[22] He concluded that 'they didn't understand our brand of peacekeeping. It involved us showing our teeth every time our authority was challenged.'[23] He did this as he believed that 'if we didn't stand firm against attack ... Gorazde might fall quite quickly. Then the killing would really begin – how many we didn't know.'[24]

---

19 Westley and Ryan, *Operation Insanity*, 4.
20 Ibid., 23.
21 Ibid., 4.
22 Westley, Interview, 2 October 2023.
23 Westley and Ryan, *Operation Insanity*, 189.
24 Ibid., 65.

The S-CALM Model

It was in this environment of Normalised Violence that the Bosnian Serbs attacked on the 28 May. They attacked the B Company observation posts (OP). OP 1 was commanded by Colour Sergeant Peter Humphreys who was an experienced soldier. The Bosnian Serbs offensive was pressed home and Colour Sergeant Humphreys was ordered to evacuate OP 1. He ordered Corporal Jones 73 to take the Saxon Armoured Personnel Carrier and he 'crashed into a BSA (Bosnian Serb Army) barricade, was engaged and returned 373 x 7.62mm rounds.'[25] Colour Sergeant Humphreys then 'withdrew in contact firing 50 x 5.56mm rounds' as he withdrew on foot he was supported by Major Westley's party who fired 'a further 1123 x 7.62mm.'[26] As he withdrew:

> He [Humphreys] led his men through the only route possible - a minefield - getting them to tread exactly where he trod. Sgt Humphreys said: "There was a piece of ground that I knew was mined. "I cleared it by running across the open ground, with the rest of the lads following in my footsteps".[27]

'Humphreys then led the rest of his team, on foot down an emergency escape route. On route, he encountered three groups of Serbs, each time he caught them by surprise.'[28] Colour Sergeant Humphreys explained that each time they surprised a group "we had guns in their heads before they knew it." 'The team disarmed the Serbs and disposed of their weapons.'[29] Major Westley explained that:

> He disarmed them because he didn't need to kill them. He said to me "we surprised them, we didn't need to kill them so I got them to lay down their weapons and we threw them over the cliff, there had been enough killing". That's the measure of Pete Humphreys, he is so cool and so collected, he can focus violence, switch the switch,

---

25 Riley, *White Dragon*, 50.
26 Ibid.
27 'Gorazde Force', The Royal Welch Fusiliers Museum Archive. https://www.rwfmuseum.org.uk/archives.html.
28 Ibid.
29 'Fusilier's battle to save Bosnians', BBC News, 5 December 2002, http://news.bbc.co.uk/1/hi/wales/2535155.stm.

he showed great moral courage, great leadership and great judgement.[30]

He concluded that Colour Sergeant Humphreys 'was the coolest man under fire I have ever seen. We had talked about when it was right to take life and when it was not. Humphreys had done his thinking.'[31]

Major Westley had prepared B Company for the hostile environment in which violence was normalised, he recalled that 'some of those young soldiers had notched up many kills on their gun belt and they had got used to taking life'. He explained that he controlled this violence by doing 'proper AARs [After Action Reviews] and went through the battle damage assessments to ensure that they understood what they did was right, but next time limit it even more.'[32]

Weak Leadership and Lack of Supervision

Weak Leadership is recognised as either *laissez-faire* or overbearing leadership and when it leads to a lack of supervision it is identified by the presence of ethical drift. To mitigate weak leadership, leaders should avoid a *laissez-faire* leadership style and retain a position of confident ethical leadership. Inertia is not an option for the ethical leader; they must be active and ready to make decisions when required. They must also remain situationally aware of their surroundings and prepared to act as required to maintain the right ethical balance in their decision-making and actions. More leadership details will be explored in chapter 10 on S-CALM Leadership.

Weak Leadership and Lack of Supervision Gorazde Case Study

Major Westley explained how 'as a UN peacekeeper, I had one arm tied behind my back but I was still going to apply my own moral code to this madness.'[33] This courageous leadership in a difficult situation was vital to the success of B Company. He considered that the commanders 'role was to make sure other people were doing the right thing and make tough decisions for them when necessary.'[34] This involved the correct level of supervision,

---

30 Westley, Interview, 2 October 2023.
31 Ibid.
32 Ibid.
33 Westley and Ryan, *Operation Insanity*, 20.
34 Ibid., 95.

but he was also keen on ensuring that he empowered others. He explained how he considered that 'mission command was absolutely relevant; I spent a lot of time talking it through and empowering my team. On training exercises, I encouraged them to make ballsy decisions if they knew my intent.'[35] He had to make some very difficult decisions and was asked about the concept of the loneliness of command. He explained that 'everyday I spoke to the soldiers, especially under the shelling. But I never felt alone as I had my Sergeant Major as my rock. I would seek his opinion and therefore I never felt isolated.'[36] Having a reliable team member that a leader can discuss options with is an important way of dealing with the loneliness of command.

Lack of Resource and Fatigue

Many operations are launched and executed with a lack of resource, especially time and workforce. A lack of time and workforce can result in those deploying being fatigued. Insufficient time prior to deployment can result in troops not being correctly trained for activities that they might have to complete. An absence of workforce can be identified by overstretch with tasks not being completed correctly. Nevertheless, many in the military believe that 'motivation and determination' allow for 'individuals to perform in real-world environments despite fatigue.'[37] Finally, when fatigue sets in it can be identified in poor cognitive functioning, poor decision-making and poor command team performance. The effect on soldiers can lead to a feeling of hopelessness and a lack of interest in the outcome of their actions.

However, many leaders believe that they do not need sleep to function. Therefore, the RMAS 2023 research survey wanted to establish firstly trainees' opinion on sleep and their concept of their ability not to be affected by a lack of sleep. The first question the trainees were asked to agree was; A commander should never sleep when their soldiers are awake? The perception of the cohort with no previous training on arrival was that leaders should be awake if their soldiers are. However, after the S-CALM model education there was 'a statistically significant shift in the average response ... suggesting that Ocdts [Officer Cadets] developed an understanding that it is acceptable for commanders to sleep.'[38] For Research question two, which was the combined

---

35 Westley, Interview, 2 October 2023.
36 Ibid.
37 Miller, Matsangas and Kenney, 'The role of Sleep in the Military', 265.
38 Vincent and Muhl-Richardson, *An Analysis of the Effectiveness of the S-CALM Model of Ethical Leadership*, 14.

answers of all trainees' 15 per cent of British cohort still believed that they should not sleep when their soldiers were awake, demonstrating that a large majority understood the need to rest when in command. However, this compared with a high 33 per cent of international trainees. The research noted that 'International Ocdts tend to value sleep less than UK Ocdts.'[39] A second question, separated in the survey from the one above, asked the trainees' to agree or disagree with the statement that; Fatigue has little effect on the way that commanders make decisions? This was aimed at their understanding that cognitive ability degrades with tiredness. The first cohort with no previous military training generally had a good understanding of the negative impacts of fatigue and lack of sleep on decision-making. For the final survey of all trainees there was a similar view of all British respondents with only 11 per cent believing that fatigue has little effect on their cognitive ability. However, in contrast 53 per cent of international trainees believed that tiredness had little effect. The research concluded that 'this difference was statistically significant ... This corresponds to the previous question on sleep, again suggesting that International Ocdts tend to underestimate the negative impacts of fatigue and lack of sleep.'[40] Despite this anomaly with the international cohort, the British responses demonstrate a good understanding of the need to rest to make ethical decisions.

When deployed, leaders can mitigate the effect of a lack of time to prepare by continuing to train whenever there is the opportunity. Operational experience can never replace the requirement for frequent training. When a leader believes that they lack the workforce to complete the tasks set to them, their immediate duty must be to prioritise the tasks to avoid overstretch and extreme fatigue. When time allows, leaders should consult with their superiors to investigate the possibility of reducing tasks allocated or increasing workforce to meet them. Fatigue is difficult to mitigate on military operations, but a leader must attempt to ensure that it does not lead to the exhaustion of their soldiers. Leaders can do this by ensuring that there is enforced rest when required. They should also consider a system of assuring that relief of individuals and/or teams is conducted to allow for recuperation. The British Army in the First World War had a system of rotating troops on the frontline to ensure that they were not exhausted. The 'trench routine was for a man and his section to spend 4 days in the front line, then 4 days in

---

39 Ibid.
40 Ibid., 15.

close reserve and finally 4 at rest.'[41] This allowed the British Army to sustain discipline and morale.

This chapter started with my recollection of getting eight hours sleep a night during a deployment to West Belfast. There are many books and articles on how to ensure that one can sleep well even in stressful situations. I am a very light sleeper and am easily awakened by the most minor issue, therefore I worked out the elements that were within my control to ensure that I could get the rest required for good cognitive ability. Below is the list of my routine conducted whenever possible to attempt to get full sleep:

- Go to bed and wake up at the same time each day. (In West Belfast I would go to bed at midnight and rise at 0800 hours)
- Exercise regularly but leave 2 to 3 hours before bed. (In West Belfast I would exercise if required in the late afternoon)
- Only take a nap at lunchtime, never in the afternoon. (I am not a great napper, but if I had been working during the night, maybe on patrol, I would nap before lunch)
- Avoid caffeine after lunchtime or 8 hours before bed. (In West Belfast I would have a can of Coke with lunch, I would then drink decaffeinated tea and occasionally coffee in the afternoon).

A leader must sleep to make good decisions, but many commanders believe that they have trained themselves to be less susceptible to fatigue. However, Jon Shay's research concludes that 'Pretending to be superhuman is very dangerous, and if leaders become 'sleepwalking zombies', from a moral point of view the adversary has done nothing fundamentally different than destroying supplies of food, water, or ammunition.'[42] Overstretch and fatigue should be avoided whenever possible as they have a discernible effect on cognitive functioning which unless dealt with can lead to unethical behaviours.

Lack of Resource and Fatigue Gorazde Case Study

The Royal Welch Fusiliers did not have enough workforce and the Battalion was both overstretched and outnumbered. Major Westley recalled that there

---

[41] The Long, Long Trail. Life in the trenches of the First World War, https://www.longlongtrail.co.uk/soldiers/a-soldiers-life-1914-1918/life-in-the-trenches-of-the-first-world-war/.
[42] Jonathan Shay, 'Ethical Standing for Commander Self-Care: The Need for Sleep,' *Parameters*, Vol. 28, No. 2 (1998), 100. 93-105.

were 'not enough people, I had a company of 120, there was another company of 120 and a third of 80. When asked to do an estimate, we calculated that it would take a division to defend Gorazde from the Serbs.'[43] In simple terms the Battalion was 'outnumbered and outgunned as the Serbs threatened to use heavy weapons and rocket-propelled grenades.'[44] Although the lack of workforce and equipment was an issue, the main lack of resource for the Battalion was that the Bosnian Serbs restricted any re-enforcements and supplies from getting into Gorazde. In the House of Commons, the MP, Dr David Clark, noted that 'they have been short of water and supplies and their lives have been pretty intolerable'.[45] On the 14 May the Commanding Officer, noted that 'today I have had to reduce rations by one-third. When soldiers are working hard, walking everywhere and carrying heavy loads because of a lack of fuel, they need calories; reducing rations is not something I do lightly.'[46] Major Westley confirmed that 'we'd already had to reduce our rations by one third.'[47] This meant that the soldiers were hungry as they continued to work. Major Westley recalled that it was hard on the solders 'when they only have a biscuit and 3 boiled sweets and you are asking them to go 20 Kilometres with 23 Kilogramme packs, we were under rationed, out gunned and out ranged.'[48] Not only did this make leadership difficult, but it also created health problems. Again, Major Westley remembered that this 'meant scurvy, it meant soldiers with teeth falling out.'[49] Nevertheless, the Battalion held together during this very testing time. Again, in the House of Commons Mr Nicolas Soames MP commented how 'the Royal Welch Fusiliers sustain themselves in their hour of trial with years and generations of experience and loyalty.'[50] Without doubt the bonding of the unit by holding to traditions and the comradeship were important to coping with this lack of resources. Even with this closeness and willingness to operate under austere

---

43 Westley, Interview, 2 October 2023.
44 BBC News, 'Fusilier's battle to save Bosnians'.
45 'Bosnia', Parliamentary Archives: British Parliamentary Papers, House of Commons Debate, Volume 260, 31 May 1995, https://hansard.parliament.uk/Commons/1995-05-31/debates/526f47de-0bc5-4417-90f7-a73fdcb93a65/Bosnia?highlight=gorazde#contribution-80e4115b-2f7a-40e4-83ae-6f3bac0619c6.
46 Riley, *White Dragon*, 42.
47 Westley and Ryan, *Operation Insanity*, 5.
48 Westley, Interview, 2 October 2023.
49 Ibid.
50 'Defence', Parliamentary Archives: British Parliamentary Papers, House of Commons Debate, Volume 264, 17 October 1995, https://hansard.parliament.uk/Commons/1995-10-17/debates/9fa3f440-5838-4818-b177-e05b13e768db/Defence?highlight=royal%20welch%20fusiliers#contribution-4484f852-2f87-4187-a28d-d95966ad31c2.

conditions, Major Westley reflected that 'when soldiers are hungry and tired, it really became wearing and it tests your leadership.'[51]

Enhanced Emotional State

In stressful situations emotions can affect the way soldiers behave. These emotions or visceral states are strong physiological influencers and cloud people's judgment as they can take precedent over rational thinking. They can be triggered by physical symptoms such as thirst and hunger and common emotions mainly consist of frustration, anger, fear, disgust and sexual arousal. To understand if officer trainees' understanding of how an enhanced emotional state causes people to behave differently than if they are not influenced by their emotions, the RMAS 2023 research survey asked the question; An enhanced emotional state leads to an increase in unethical behaviour? Trainees with no previous military experience understood that an enhanced emotional state influences behaviour and this understanding increased following education in the S-CALM model. In the final survey 79 per cent of British trainees agreed that emotions could lead to unethical behaviour whilst five per cent believed that they did not. In the international cohort 67 per cent believed that emotions made a difference to behaviour. Therefore, generally most trainees understood that they needed to monitor their emotional state so as to make ethical decisions.

A key mitigation technique for enhanced emotions is the Combat Technical Breathing exercise that has already been mentioned. Another way to try and reduce the effect of negative emotions is to try and improve the emotion state and strengthen positive emotions. An experiment was conducted in 1972 to establish if a person who was 'feeling good' would spontaneously help to pick up papers that were dropped in front of them when it had been proved that most people would not help. In this experiment the 'feel good' factor was making a free phone call from a telephone box when they found a dime after the call. Of those that paid for the call only 4 per cent assisted a person who they bumped into to pick up their papers. However, in the group that found the dime 87.5 per cent assisted in picking up the papers due to the 'feel good' factor. The experiments 'finding provides evidence for the "warm glow" hypothesis' in that 'people who feel good themselves are

---

51 Westley, Interview, 2 October 2023.

more likely to help others.'[52] Therefore, establishing a positive environment should lead to positive actions. A final significant skill is the application of emotional intelligence in leadership. Emotional intelligence in leadership has been described has having five key aspects for assessing how leaders cope with stress, they are:

- Self-Awareness: Understanding one's internal state, resources and intuition
- Self-Regulation: Managing one's internal state and impulses
- Self-Motivation: Having the emotional ability to reach one's goals
- Empathy: Being aware of others' feelings, needs and concerns
- Social Skills: Inducing desirable responses from others.[53]

Displaying emotional intelligence can allow leaders to control their emotions. US Army doctrine notes that 'effective leaders control their emotions. Emotional self-control, balance, and stability enable leaders to make sound, ethical decisions.'[54] A great exemplar of this was Admiral James Stockdale. On 9 September 1965, Stockdale ejected from his plane having been hit by enemy fire over North Vietnam. He was held as a prisoner of war in the Hỏa Lò Prison, known as the Hanoi Hilton, for seven and a half years. He was the senior naval officer in the prison and was routinely tortured. A key emotion that can lead to poor ethical behaviour is fear. Research has shown that 'fear is an emotion, and controlling your emotions can be empowering.'[55] On release he recalled that 'I whispered a "chant" to myself as I was marched at gunpoint to my daily interrogation: "control fear, control guilt, control fear, control guilt."'[56] He also remembered that he deflected his 'gaze to obscure such fear or guilt as doubtless emerged in my eyes when I temporarily lost control under questioning.'[57] Major Westley's display of emotional intelligence in leadership was the key component in surviving and inspiring others whilst in the most extreme conditions in Gorazde.

---

[52] Alice Isen and Paula Levin, 'Effect of feeling good on helping: Cookies and kindness,' *Journal of Personality and Social Psychology*, Vol. 21, No. 3 (1972), 386.
[53] Shanthakumary Aloysius, *'The Role of Emotional Intelligence in Leadership Effectiveness'*, Conference Paper, Jaffna University Research Conference, October 2010, 4.
[54] US Army, *ADP 6-22*, 4-3.
[55] James Stockdale, *'The Stoic Warrior Triad: Tranquillity, Fearlessness and Freedom'*, A lecture to the student body of The Marine Amphibious Warfare School, Quantico, Virginia, 1995.
[56] James Stockdale, *'Courage Under Fire: Testing Epictetus's Doctrines in a Laboratory of Human Behavior'*, Hoover Essays No. 6, Stanford University, 1963, 14.
[57] Ibid.

Enhanced Emotion State Gorazde Case Study

The Royal Welch Fusiliers had been exposed to some horrific experiences and there is little doubt that this would have had an emotional effect on the soldiers involved. Major Westley recalled that he 'was worried about the morale, as well as moral, state of my young soldiers. They had seen and done things that would remain with them for life.'[58] Therefore he had to monitor the emotion of the B Company soldiers. To help to do this he described how 'I always tried to find 30 mins or an hour a day to walk, think, horizon scan, mainly alone but sometimes with other people, it was my way of regulating it.'[59] He explained that he did this as he:

> didn't want to get angry or disconsolate as I wanted to set the example. Walking allowed me to stay grounded, especially if I had seen some that made me really sick or angry that day, like a child being killed, you have to bottle that and if you can't regulate it, you put it in a box and if you can't regulate it you need to open the box one day. However, you do it doesn't matter, but as long as the blokes don't see you starting to lose it'.[60]

With this daily walking he believed he inspired others and that 'despite the tension I was feeling inside, I was casually strolling around, trying to look relaxed.'[61] This was Major Westley's way of emotional regulation and importantly demonstrating to the solders of B Company that their leader was in control. He went on to explain that 'there is an element of acting when being an officer, there are times when you absolutely want to show your emotions to get your point across and there are times when you don't want to show them.'[62] Major Westley's self-regulation worked well during the operation and he explained that even when they finally extracted from Bosnia and were safe that he couldn't show any happiness or relief, he questioned 'why did I feel strangely detached from this happy scene? I couldn't let anything out, because where would it stop?'[63]

---

58 Westley, Interview, 2 October 2023.
59 Ibid.
60 Ibid.
61 Westley and Ryan, *Operation Insanity*, 285.
62 Westley, Interview, 2 October 2023.
63 Westley and Ryan, *Operation Insanity*, 290.

Summary

In summarising the deployment to Gorazde, Major Westley explained how 'we had to plan for when and not if the Serbs were going to attack.'[64] He concluded that 'had we not been there on May 28, Gorazde would have suffered the same fate as Srebrenica.'[65] 'The Dutch, in particular, seemed gullible and they were caught out.'[66] In the House of Commons Mr Nicolas Soames commented on how the world 'saw the steadiness and courage displayed by the Royal Welch Fusiliers.'[67] The UN Commander in Bosnia, Lieutenant General Rupert Smith, said of the Battalion that 'they showed, in ample measure, all that is best in the British Army; professionalism, courage, cheerfulness and a readiness to overcome any obstacle.'[68] For 'personal example and leadership over a prolonged period' Major Westley was awarded the Military Cross.[69] In addition, Colour Sergeant Humphreys became only the second British soldier ever to be awarded the Conspicuous Gallantry Cross for his actions. His citation stated that he had 'tremendous presence of mind, aggressive spirit and coolness under fire.'[70]

It has been demonstrated above that situational influencers can be enhanced if the system heightens the problem by not applying any checks to their development due to weak leadership or supervision. However, these situational influencers do not have to lead to the development of common cognitive behaviours if a leader can apply accountability, leadership and has a good moral compass.

---

64 Westley, Interview, 2 October 2023.
65 BBC News, 'Fusilier's battle to save Bosnians'.
66 Westley and Ryan, *Operation Insanity*, 174.
67 Parliamentary Archives, 'House of Commons Debate Defence'.
68 Riley, *White Dragon*, 5.
69 Westley and Ryan, *Operation Insanity*, 303.
70 Ibid.

7

S-CALM:
C – Common Behaviour (Individual)

Introduction

I was commissioned in 1983 and experienced my first real pressure to conform to fit in as a young Army officer. When I joined my battalion, the junior officers were members of what they called the 'Zoo Club'. To be in the 'Zoo Club' you had to have a small tattoo of an animal on your bottom. That does not sound radical today, but in the 1980s British Army officers simply did not have tattoos. It was therefore seen to be rebellious to have this hidden tattoo. I was placed under considerable peer pressure to have one and I really wanted to fit into my new group and be a member of the club. However, despite this I resisted the pressure to conform. Once I had made my stand and did not conform, the peer pressure by those that were enforcing the 'Zoo Club' on new subalterns was broken and as new officers joined and refused to be tattooed, the club slowly disappeared.

Once the power of the situation, enhanced by the situational influencers, take effect people begin to use 12 common behaviours. The next important element of applying the S-CALM model is to recognise and mitigate these common behaviours. Recognising these behaviours and how to mitigate them is a fundamental skill for an ethical leader. Ethical leaders must understand and be able to recognise the 12 common cognitive behaviours in both themselves and their teams and then mitigate their effects. To achieve this understanding leaders must know themselves well. If leaders can identify that they are already susceptible to a certain behaviour, they can be prepared to mitigate for it when under pressure caused by situational influencers.

This flow of unethical action has been demonstrated and a 'large body of research in social psychology seems to show that human behavior [sic] is explained to a surprising extent by the external situational factors (such as stress, peer pressure, or orders from authority figures) to which a person is exposed.'[1] Leaders also need to understand their teams. This allows them to set the narrative of what is and is not acceptable behaviour, to identify the six common group behaviours and prepares them to steer their teams away from unethical actions.

Although militaries have been slow to fully understand how the flow of unethical action takes place, there is some recognition of the fact that individual and team behaviours contribute to unethical actions. British Army doctrine states that 'leaders who set a bad example are dangerous, as inappropriate behaviour can be infectious and can quickly become the norm if not corrected.'[2] Although this statement is obvious, it does identify that a leader must be an exemplar in their behaviours and that they have influence over their team, an idea that will be explored more in chapter 10. The US Army doctrine explains how 'all Army members who witness these behaviors [sic] have a responsibility to prevent, intervene, counter, or mitigate them.'[3] This identifies the need to for everyone to have the confidence to intervene and not be a bystander, which will be considered more later in this chapter. US SOF in their *A Special Operations Forces Ethics Field Guide* explain how 'upholding good ethical behaviors [sic] takes work, and you must train for it.'[4] However, it falls short of explaining how this will be done. The best doctrine statement in this area comes from the Australian Defence Force which explains that the 'Ethical Decision-making Framework asks us to examine our biases and stereotypes, as well as the effect they may have on our ethical decision-making.'[5] It is this examination of the effect of the identified common behaviours and how they can be positively handled under stress which is the basis of this chapter.

---

1 Wolfendale, *The Case of War Crimes*, 276.
2 MOD, *Army Leadership Doctrine*, 2-9.
3 US Army, *ADP 6-22*, 8-8.
4 US Army, *A Special Operations Forces Ethics Field Guide*, United States Special Operations Command, 2023, I.
5 ADF, *Military Ethics*, 31.

S-CALM Common Individual Behaviours and Case Studies

In this chapter each of the six common behaviours that are mainly focused on individuals will be studied with the aim of recognising what the behaviour might look like in a stressful situation. The main mitigation of common behaviours in the S-CALM model is achieved through the application of accountability, leadership and the moral compass, which will all be explored in later chapters. Nevertheless, where appropriate some mitigating ideas will be suggested. There will also be a positive case study example for each of the behaviours demonstrating a situation when an individual or team overcame a behaviour. One of the background concepts at the start of this book was the 'Banality of Evil', where everyone can do unethical actions. The case studies in this chapter are grounded in the principal of the 'Banality of Good', in that everyone is also capable of conducting ethical actions given the right conditions and the confidence to resist their initial instinct.

Social Comparison Theory and Conformity

As explored in chapter 3, social comparison theory and conformity impose a strong obligation on people to imitate their teams thinking and actions. In his original work, Festinger explained how 'people tend to move into groups which, in their judgement, hold opinions which agree with their own and whose abilities are near their own.'[6] These groups of likeminded people in militaries are often referred to as 'cliques' and will be explored more when groupthink is investigated later. However, the power of the obligation to conform with others was termed as social proof by Robert Cialdini in *Influence: Science of Persuasion*. In this book Cialdini explains how 'social proof is a phenomenon where people follow and copy the actions of others in order to display accepted or correct behavior, [sic] based on the idea of normative social influence.'[7] Militaries thrive on the concept of normative social influence where people conform to be accepted as part of a team. Nonetheless relating to Festinger's work, when teams become cliques, they can become liable to groupthink and possibly othering of those who are not included. In these situations, leaders need to ensure that they allow diversity within their teams and encourage individuals to challenge and speak out in groups.

---

6 Festinger, 'Theory of Social Comparison Processes', 136.
7 'Is social proof really that important? Here's how to use it', Shanelle Mullin, 30 August 2023, https://cxl.com/blog/is-social-proof-really-that-important/#h-what-is-social-proof.

How to make and receive a challenge will be explained more in chapter 10. Another important aspect of social comparison theory and conformity is the need to act quickly under peer pressure. This can be mitigated by allowing time to think when faced with a decision. The STOP Protocol, in chapter 12, clarifies how this thinking time can be achieved.

Social Comparison Theory Case Study – Crew of HMS Coventry 1982

In the spring of 1982 HMS *Coventry*, a Type 42 destroyer of the Royal Navy was on Exercise Springtrain 82 in the Mediterranean Sea. *Coventry* was commanded by Captain David Hart Dyke and had a crew of 28 officers and 271 ratings. *Coventry* was described as 'a confident ship; men were quietly confident in their own ability to fight.'[8] On the 2 April she was warned to move to Ascension Island to join other ships who were going to the South Atlantic to take part in the Falklands War. On the way to Ascension Island *Coventry* was paired with HMS *Aurora*, a ship heading back to the UK, which was to pass on all available stores and equipment. Captain Hart Dyke explained how 'it was quite traumatic when we knew where we were going to continue on south from Ascension. Preparing a ship for war focuses on the dangers ahead. Anxious letters from home, increased our concerns as well. It was a very unsettling time.'[9] Captain Hart Dyke made the decision to ask the crew members of Coventry if any of them wished to return to the UK; he said that he would replace any 'who had strong reasons to get back home.'[10] The crew had conducted Exercise Springtrain in the Mediterranean together and Captain Hart Dyke believed that they were a close and confident team and there would be great social pressure not to leave their comrades and to continue to the Falklands. However, Captain Hart Dyke created an environment where sailors felt empowered and had the self-confidence to speak up. Captain Hart Dyke recalled that he 'had five sailors on board who had family concerns back home or important events planned.'[11] Therefore 'five of the crew were duly swapped with similarly ranked seamen from

---

8 MOD, *Board of Inquiry – Report into the Loss of HMS Coventry*. (London: MOD, 1982), 6.
9 'Falklands 40: David Hart Dyke', SSAFA. https://www.ssafa.org.uk/support-us/our-national-campaigns/falklands-40/falklands-40-david-hart-dyke.
10 Ibid.
11 David Hart Dyke, *Four Weeks in May: A Captain's Story of War at Sea*. (London: Atlantic Books, 2008), 20.

another ship that was heading back to the UK.'[12] The record confirms that '5 members of Coventry's crew were released to return home on Aurora, and 5 members of Aurora's crew replaced them.'[13] On 27 April *Coventry* entered the Total Exclusion Zone, a 200-mile cordon around the Falkland Islands as an air defence ship that was part of the protection for the British aircraft carriers. Captain Hart Dyke recalled that 'when we knew we were actually going to start fighting we faced it bravely. And morale rose to a high point. Of course, it required strong leadership, discipline and good training to be confident in what we would be facing.'[14] The exchange of personnel at Ascension Island demonstrates how given the right environment people can overcome the strong social pressure that is created by social comparison theory.

Conformity Case Study – Batang Kali Massacre

The Batang Kali massacre, in which the Scots Guards killed 24 innocent men during the Malaya Emergency has already been outlined. However, not all soldiers present took part in the massacre. One of those who took part in the massacre explained how, 'Sergeant Hughes addressed the patrol. The men and boys must be shot, he told them. Anyone too 'squeamish' should take one pace forward and fall out. Alan Tuppen stayed put. James Fern and Victor Remedios both opted out and Hughes ordered them to stand guard on the road.'[15] Another participant confirmed that 'the sergeant in charge informed the patrol what was going to happen. Those who were not prepared to participate should say so. On hearing that, Remedios and another soldier had opted to guard the women and children in the truck rather than participate in the shooting.'[16] Guardsman Victor Remedios recalled 'they had been instructed that those opposed to the killing plan could fall out.' He was ordered to go and guard the women and children and therefore 'found I need play no part in the killing. I went up the road to where the lorry with

---

12 'David Hart Dyke, the captain of HMS Coventry recalls the horror of his ship sinking in the Falklands War: 'It was black, with people on fire', The Telegraph, 18 May 2012, https://www.telegraph.co.uk/news/worldnews/southamerica/falklandislands/9272406/David-Hart-Dyke-the-captain-of-HMS-Coventry-recalls-the-horror-of-his-ship-sinking-in-the-Falklands-War-It-was-black-with-people-on-fire....html.
13 'HMS Coventry D118', The Falklands War, https://www.hmscoventry.co.uk/d118/falklands/.
14 SSAFA, 'Falklands 40: David Hart Dyke'.
15 Hale, *Massacre in Malaya*, 205.
16 Ward and Miraflor, Slaughter and Deception at Batang Kali, 95.

the women was parked and acted as a guard.'[17] Despite a participant in the massacre saying they took part because of the discipline in the Guards, an opportunity was given to the patrol by Sergeant Huges not to take part when he offered the chance to fall out. Guardsmen Fern and Remedios had the ability to overcome the strong social obligation to conform and went against their social group by accepting the offer to fall out and therefore refusing to take part in the massacre.

Deindividuation

Deindividuation envisages people giving up their responsibility, especially when they believe that they are anonymous within a group, which is the norm for junior ranks within the majority of militaries. This allows individuals to conduct acts that they would not consider to be acceptable if they were alone, but as part of the team they suffer the diffusion of responsibility and pass any feelings of guilt or shame on to the team or the leader. The fundamental requirement to counter this behaviour is the need for individuals to be held accountable for their actions. Firstly, leaders need to gain self-awareness as this is 'highly associated with leadership effectiveness.'[18] A leader who is self-aware understands their own thoughts, feelings, values and actions. Therefore, the self-aware leader is comfortable with their own actions and can concentrate on their team. This allows them to focus on their team members as individuals and empower their followers to consider their actions and not feel anonymous. The leader is then able to hold individuals to account for their actions within the team. Accountability is explained and explored in depth in chapter 9. Individual team members are thus empowered to make their own decisions understanding that there is no group indemnity.

Deindividuation Case Study – Sergeant Joe Darby at Abu Gharib

The abuses, mainly committed by Charles Garner and Lynndie England at Abu Gharib prison in 2004 have already been outlined in chapter 4. However, this case study refers to Sergeant Joseph Darby who was the whistleblower in the Abu Ghraib prisoner abuse case. Joe Darby had witnessed the abuses and was uneasy with what he saw. He recalled that 'when I was in Iraq, I had a very difficult decision to make … It just seemed like the right thing to do at

---

17 Ibid., 88/89.
18 Aloysius, 'The Role of Emotional Intelligence in Leadership Effectiveness', 6.

the time.'[19] Joe Darby encouraged Charles Garner to hand over copies of the photographs of the abuses. He then struggled with what to do next as he had known Lynndie England since basic training. He recalled that 'I was faced with the toughest decision. On one hand, I had my morals and the morals of my country. On the other, I had my comrades, my brothers in arms.'[20] After 3 weeks of deliberation he sent the photographs to Special Agent Tyler Pieron of the US Army Criminal Investigation Command who was stationed at Abu Ghraib Prison, which initiated the investigation and subsequent trials. One of the trial judges, John Shattuck, said that 'it took courage for Darby to stand up for justice. He must have known that what he did would make him a pariah with his colleagues, but he followed his conscience.'[21] Joe Darby is a great example of an individual that was aware of wrongdoing by the team and instead of allowing deindividuation to induce inertia, retained their own individual moral code which dictated their action.

Obedience

In chapter 3 it was explained how obedience is the norm in the military as personnel are expected to follow orders instantly on the battlefield with no hesitation. However, it also noted that obedience to orders was one of the common behaviours that led to abuse. The British Army has a long heritage of questioning dubious orders, going back to the end of the Second World War and what was known as the Nuremberg Defence, in which war criminals said that they were 'only following orders'. However, the London Charter of the International Military Tribunal, which set up the Nuremberg Trials stated that 'the fact that the Defendant acted pursuant to order of his Government or of a superior shall not free him from responsibility under international law.'[22] British Army doctrine still clearly states that 'a subordinate who judges an order to be immoral, unethical, or illegal has an obligation to refuse to execute it.'[23] US Army doctrine is similar, but not as clear, when it states that 'sometimes the situation requires a leader to stand firm and disagree with

---

19 'Remarks made by Sergeant Joseph Darby on accepting the 2005 Profile in Courage Award' John F. Kennedy Courage Awards 2005, 16 May 2005, https://www.jfklibrary.org/events-and-awards/profile-in-courage-award/award-recipients/joseph-darby-2005.
20 Ibid.
21 Ibid.
22 'Agreement for the prosecution and punishment of the major war criminals of the European Axis. Signed in London, on 8 August 1945', UN, 1951. https://www.un.org/en/genocideprevention/documents/atrocity-crimes/Doc.2_Charter%20of%20IMT%201945.pdf.
23 MOD, *Army Leadership*, 3-5.

a supervisor on ethical grounds. These occasions test one's character and moral courage.'[24] To assist with the ability to challenge and be challenged, leaders are advised to nominate 'at least one individual with the necessary prestige and confidence to challenge the commander, when appropriate.'[25] Therefore a leader must be open to legitimate challenge. To further support the ability of team members to challenge a leader, it is recommended that whenever possible a leader uses a participative leadership style. The leader using the participative style asks the question of the team 'what do you think?' British Army doctrine describes participative leadership as a style in which 'the leader asks for and values input from the team. They create an environment where ideas and timely constructive criticism is welcome, building commitment through participation.'[26]

The RMAS 2023 research survey asked a question to gauge the trainees' interpretation of obedience. The question asked: If ordered would you restrict a prisoners sleep to gain critical information? For those that had no previous training 'only 34% of the cohort was ready to say no and 23% were unsure what to do' on arrival.[27] By the end of their year, only four per cent of this cohort along with all other British trainees stated that they would obey the order to restrict sleep. This stand is stronger than just offering a challenge and is known as intelligent disobedience. As with challenge, the British Army has a long heritage and legal obligation to refuse an unlawful order. Intelligent disobedience is enshrined as the right to disobey orders that will lead to illegal actions, UK doctrine describes it as 'the actions of a follower who judges an order to be immoral, unethical, or unlawful.'[28] In a piece on intelligent disobedience the Centre of Army Leadership's Professor Lloyd Clark stated that 'a healthy army with the appropriate culture might be especially well-placed to inspire intelligent disobedience because of its esprit de corps and consistent focus on professional development.'[29] The key element of both challenge and intelligent disobedience is the appropriate culture of mutual respect that a leader builds with their team. Reasonable challenge will be dealt with in depth and intelligent disobedience will be recapped in chapter 10

---

24 US Army, *Army Leadership*, 2-8.
25 Navy Command, Telemeter-Internal Review, C3.
26 MOD, *Army Leadership*, 5-10.
27 Vincent and Muhl-Richardson, *An Analysis of the Effectiveness of the S-CALM Model of Ethical Leadership*, 15.
28 MOD, *British Army Followership Doctrine*, 16.
29 Lloyd Clark, *Leadership Insight No.1 March 2017: The Intelligently Disobedient Soldier*. (Camberley: The Centre for Army Leadership, 2017), 2.

Obedience Case Study – General Jackson and the Pristina Airport Incident

In June 1999 a NATO force, known as KFOR, began to enter Kosovo to bring peace to the war-scarred province under Operation Allied Force. The overall commander of the operation was US Four Star, General Wesley Clark, the Supreme Allied Commander Europe (SACEUR). The commander of the Allied Rapid Reaction Corp on the ground was Lieutenant General Mike Jackson of the British Army. Although there were a few initial frictions with the plan for Operation Allied Force, on 12 June orders were given for the ground troops under General Jackson to start moving into Kosovo. However, at this point it became clear that the Russian Government, which had expected to be invited to be part of the ground force, had dispatched a 200 strong force from Bosnia to Pristina airport in order to secure it and fly more troops into the city before NATO troops arrived. In this situation, General Clark ordered General Jackson to deploy NATO troops to the airport by helicopter in order to stop the Russians from seizing it. At the time General Jackson told the BBC there was the possibility 'of confrontation with the Russian contingent which seemed to me probably not the right way to start off a relationship with Russians who were going to become part of my command.'[30] He recalled after the event that if the 'military judgement was so strongly against this operation I should be prepared to oppose it publicly, even if it meant putting my own career on the line.'[31] He also reflected that 'for the first time in my almost forty years in the army I had been given an order that I felt I could not in principle accept. In a few minutes' time my career could be over.'[32] When General Clark called and ordered General Jackson to deploy troops to the airport he replied 'Sir, I am not going to start World War Three for you.'[33] General Clark then said 'Mike, do you understand that as a NATO commander I'm giving you a legal order, and if you don't accept that order you'll have to resign your position and get out of the chain of command?' Jackson replied 'SACEUR, I do'.' General Clark then concluded 'OK. I'm giving you an order to block the runways at Pristina airfield. I want it done. Is that clear?'[34] General Jackson still refused. *The Guardian* newspaper reported how 'both generals turned to their political masters for support, but while the British government backed

---

30 'Confrontation over Pristina Airport', BBC News, 9 March 2000, http://news.bbc.co.uk/1/hi/world/europe/671495.stm.
31 Jackson, *Solider*, 261.
32 Ibid., 262.
33 Ibid., 272.
34 Ibid., 273.

General Jackson's judgment, General Clark received no support, effectively meaning his orders were overruled.'[35] The order was not carried out and after some negotiation the Russian troops became part of KFOR. This case study demonstrates a clear example of a leader who was prepared to carry out intelligent disobedience in the face of an order that, although legal, was ill advised and could have escalated an already tense situation.

Cognitive Dissonance Theory

When people have cognitive dissonance, they are in a state of confliction. They are aware of what they ought to do but conduct the actions that they want to do. This causes them the mental dissonance or conflict. Most reasonable people will recognise when they are in a state of cognitive dissonance and that their actions are not aligned with their beliefs. However, in this state people will try to justify their action, such as blaming the victim or try to legitimise the actions by moralising what they are doing. In many of the case studies explored in chapters 3 and 4, those conducting unethical actions had convinced themselves that their actions were morally legitimate and that they were doing their actions for the right reasons. The leader can mitigate the effect of cognitive dissonance by firstly ensuring that their actions are compatible with their moral compass, which will be explored in detail in chapter 11. They must also ensure that there is the psychological safety within the team to question decisions and actions they feel may not be correct, but which the team is attempting to incorrectly rationalise.

Cognitive Dissonance Theory Case Study – Sergeant Franceschi at Dien Bien Phu

The Battle of Dien Bien Phu was the last big battle of the French Army in French Indochina, now Vietnam, in 1954. The French had committed a huge force to a series of strong points in the Dien Bien Phu valley with the idea of bringing the Viet Minh to battle. The French had underestimated the ability of their enemy to concentrate forces, especially artillery and they soon found themselves surrounded and cut off from external assistance. Sergeant Paul Franceschi was part of 2nd Company, 8th Battalion Colonial Parachute Regiment during the Battle of Dien Bien Phu. On 31 March 1954, he took part in a desperate fight to recapture the hilltop strong point Dominque II

---

35 "I'm not going to start Third World War for you", *The Guardian*, 2 August 1999. https://www.theguardian.com/world/1999/aug/02/balkans3.

which had fallen to the Viet Minh. When he reached the summit, he head for the mortar emplacement, with the idea of getting the mortars back in action and firing at the retreating Vietnamese troops. As he approached the mortar emplacement he saw a wounded enemy soldier, whose automatic pistol and map case suggested a non-commissioned officer. Sergeant Franceschi recalled that the wounded soldier was 'a can-bo, or political commissar, the worst type of Vietminh.'[36] There was then a conversation with another soldier Private First Class Froissard. Private Froissard asked 'Do we finish him off Sergeant?' to which the Sergeant replied 'No! He is wounded, and we must do the same for him as for anyone else who is wounded. Put a dressing on him.' Private Froissard replied 'Even when we haven't enough as it is?' and he was told 'Do what I tell you! There are enough dead around us for you to find one.'[37] The dressing was applied. Sergeant Franceschi took the wounded soldier's map case and weapon and 2nd Company were ordered to withdraw from the hill before the Viet Minh counter attacked. Despite the urge not to help the enemy, especially as he was a political commissar, in this desperate situation Sergeant Franceschi did what he ought to have done and not what Private Froissard wanted to do. There is an interesting follow up to this case study. In late 1954 when Sergeant Franceschi was a prisoner of war and very ill, the wounded Viet Minh political commissar recognised Franceschi and ensured that he was cared for and subsequently survived captivity.

Bystander Effect

The social obligation to conform or the influence of groupthink not to speak out can cause individuals to suffer the diffusion of responsibility and ultimately the bystander effect. The opposite of the bystander effect is bystander intervention. Bystander intervention 'occurs when an individual breaks out of the role of a bystander and helps another person in an emergency.'[38] Bystander intervention can take place when an outsider from the group notices that the group in not acting correctly. Edmund Burke, the British 18th Century statesman, is attributed with saying 'All that is necessary for the triumph of evil is for good men to do nothing'. It is known that he did write in 1770 in a political pamphlet called *Thoughts on the Cause of the*

---

36 Royal Benoit, *The Ethical Challenges of the Soldier: The French Experience*. (Paris: Economica, 2012), 59.
37 Ibid.
38 Hogg and Vaughan, *Social Psychology*, 657.

Present Discontents in which he stated that 'when bad men combine, the good must associate; else they will fall, one by one, an unpitied sacrifice in a contemptible struggle', but this is not so catchy. The philosopher John Stuart Mill gave a speech in 1867 at the University of St Andrews that 'Bad men need nothing more to compass their ends, than that good men should look on and do nothing'. In October 1916 it was reported that The Reverand Charles Aked used the following in a speech 'It has been said that for evil men to accomplish their purpose it is only necessary that good men should do nothing.'[39] This is the nearest to the phrase which became famous when it was said by President John F. Kennedy in 1961, who credited it to Burke. The important point is that the need for a person to step in and take action when they identify wrongdoing has been a sentiment for over 300 years. It was reiterated by Lieutenant General David Morrison, Chief of the Australian Army, who summarised this idea when he stated that 'the standard that you walk by is the standard that you are prepared to accept.'[40]

The RMAS 2023 research survey wanted to understand trainees' knowledge of the need to intervene when they witnessed unacceptable behaviour, even if the perpetrators were not under their command. The initial response by those with no previous training was 97 per cent positive that they would take action and this remained constant in all three surveys. In the final survey for all respondents 'of UK Ocdts [Officer Cadets], 98% agreed that you should take action to address poor behaviour outside of your platoon and of international Ocdts, 87% agreed.'[41] The results demonstrate a good understanding of the need for a leader to intervene and not be a bystander. Although it can be difficult, it is important that leaders at every level have the confidence to intervene. In the Mai Lai massacre, it was noticed of C Company 'that relations between officers and soldiers were much more informal and personal than in other companies.'[42] In this situation the leaders did not have the psychological separation which would have allowed them to conduct bystander intervention when required to act. This separation is termed the loneliness of command and will be explored in depth in chapter 10.

---

[39] 'The Only Thing Necessary for the Triumph of Evil is that Good Men Do Nothing', Quote Investigator, 2010, https://quoteinvestigator.com/2010/12/04/good-men-do/.
[40] David Morrison, 'The standard you walk past is the standard you accept', ADF investigation, 2013, https://speakola.com/ideas/david-morrison-adf-investigation-2013.
[41] Vincent and Muhl-Richardson, *An Analysis of the Effectiveness of the S-CALM Model of Ethical Leadership*, 16.
[42] Tony Raimondo, 'The My Lai Massacre: A Case Study', Human Rights Program, School of the Americas, Fort Benning.

The British MOD has a system called the Four Ds which is a structure designed to assist people conduct bystander intervention in difficult situations. The introduction to the four Ds states that 'if you think something is unacceptable, you'll probably find other people do too. So if you intervene you will more likely be joined by others', it goes on to explain that there are 'four methods of positive intervention; Direct, Distract, Delegate and Delay.'[43] In summary the four methods are as follows:

- *Direct:* If people have confidence they should directly intervene
- *Distract:* If people don't feel they can directly intervene they can shift the focus from what is happening and interrupt the behaviour to create a deflection
- *Delegate:* If people don't feel that it is safe or that they are not confident enough to intervene, they can find someone who is confident enough to do so
- *Delay:* If people can't intervene in the moment, they can offer support to the victim afterwards
- *There is a fifth D, Document*: Documenting the incident can be useful, especially if people are going to report what happened.

The Four Ds give people options for ensuring that there is never a case where unacceptable behaviour cannot be stopped from spiralling into more serious unethical actions.

Bystander Effect Case Study – Warrant Officer Thompson at My Lai

The My Lai massacre in which up to 500 innocent Vietnamese people were killed was outlined in chapter 4 however this case study will investigate how it was stopped. Above the village during the operation was a reconnaissance helicopter flown by Warrant Officer Hugh Thompson with two door gunners, Specialists Glenn Andreotta and Lawrence Colburn. As the ground troops advanced WO Thompson observed them appearing to shoot civilians and he sent a message to other helicopters, knowing it would be monitored at headquarters saying 'It looks to me like there's an awful lot of unnecessary killing going on down there. Something ain't right about this. There's bodies

---

[43] MOD, *Active Bystander Fundamentals: Alternative Format* (MOD, 2023), 15.

everywhere. There's a ditch full of bodies that we saw.'[44] He landed the helicopter and talked to Lieutenant Calley, who told WO Thompson to stay out of it. However, when WO Thompson saw troops pushing civilians into a bunker, he believed that they were about to kill them and he landed the helicopter again. As they came down he said to the crew, 'Y'all cover me! If these bastards open up on me or these people, you open up on them.'[45] He took control of the situation and brought in troop carrying helicopters to extract the civilians. He then radioed the headquarters to stop the attack as it was a massacre. WO Thompson gave evidence at Lieutenant Cally's courts martial and was ostracized by the Army for some time. However, in 1998, 30 years after the event he and Specialists Andreotta and Colburn were awarded the Soldier's Medal, the highest US award out of combat. WO Thompson's citation read that it was 'for heroism above and beyond the call of duty on 16 March 1968, while saving the lives of at least 10 Vietnamese civilians during the unlawful massacre of noncombatants by American forces at My Lai'. It went on, 'Warrant Officer Thompson's Heroism exemplifies the highest standards of personal courage and ethical conduct, reflecting distinct credit on him, and the United States Army.'[46] When asked at the award ceremony why he did it WO Thompson said 'I saved the people because I wasn't taught to murder and kill.'[47] After the award WO Thompson gave many lectures on My Lai and in one recalled that 'if you don't think it's right, it more than likely is not right.'[48] The difference at My Lai was that he had the moral courage and confidence to directly intervene and stop the massacre when so many others applied the diffusion of responsibility and did nothing.

Status Quo Bias

No organisation operates in a vacuum and is therefore constantly subject to change. General Eric Shinseki, when Chief of Staff for the US Army, initiated the Army Transformation Campaign to address the emerging strategic challenges of the early 21st Century. When faced with critics he said, 'if you

---

44 Warrant Officer Historical Foundation, The Forgotten Hero of My Lai: The Hugh Thompson Story, https://warrantofficerhistory.org/PDF/Forgotten_Hero_of_My_Lai-WO_Hugh_Thompson.pdf.
45 Ibid.
46 Hugh Thompson, *Moral Courage In Combat: The My Lai Story*. Lecture to Center for the Study of Professional Military Ethics, 2003, 4.
47 Thompson, *Moral Courage In Combat*, 7.
48 Ibid., 16.

The S-CALM Model

Capability Barriers	Motivational Barriers	Opportunity Barriers
Do I understand it? Do I know how to do it?	Do I believe I can do it?	Does the environment make it difficult or impossible?
Do I have the physical ability to do it?	Will it lead to a positive or negative outcome?	Do I have the resources and the time needed to do it?
Will it capture and hold my attention?	How do I feel about doing it?	What role models in my environment will encourage me to do it?
Will I be able to evaluate the different options and make the right decision?	Have I got a clear goal or target? Is the goal a priority for me? Who will hold me accountable?	Is it the norm in my social group to do it? Will I be perceived negatively if I do it? How do my peers influence my behaviour?

Table 7.1   Barriers to Change

don't like change, you're going to like irrelevance even less.'[49] As explored in chapter 3, militaries can be keen to cling to traditions and tend to put up three main barriers to any transformational change centred around capabilities, motivation and opportunity. Based on these three areas from the MOD's *Behavioural Science in Defence: A toolkit for practitioners*, people resisting change tend to ask questions such as those in Table 7.1.

In many cases these questions are raised because people want to understand the change and a leader must be prepared to explain to their team why it is required. When considering this narrative, a leader should try and ensure that their explanation is clear and simple, applies to the thinking of the team and is concise. The leader must also be prepared to be challenged, as not all change will be considered as ethically sound. How to give a challenge to a superior and accept challenges from followers will be explained in detail in chapter 10 on leadership.

Status Quo Bias Case Study – Police Battalion 101

The way that Police Battalion 101 turned from ordinary men into mass murders was explained in chapter 4. This case study will investigate

---

[49] Peter Boyer, A Different War, *The New Yorker*, 23 June 2002, https://www.newyorker.com/magazine/2002/07/01/a-different-war.

why most men became killers, 'while a minority of perhaps 10 percent – certainly no more than 20 percent – did not?'[50] In a unit that was operating in hostile territory it has been noted that 'those who did not shoot risked isolation, rejection, and ostracism – a very uncomfortable prospect within the framework of a tight-knit unit.'[51] However, it has been shown that a sizable majority chose not to shoot. This resistance to the status quo appeared to have started early and many 'testimonies are filled with stories of men who disobeyed standing orders during the ghetto-clearing operations and did not shoot infants or those attempting to hide or escape.'[52] This opposition to the norm also included one of the officers and it was recorded that 'Lieutenant Buchmann did not kill.'[53] It is possible that due to their large numbers and having officers amongst them this minority group were able to resist the need to conform and were able to obey their conscience and not kill innocents. It does give rise to an interesting question about those that chose to observe the status quo and killed in these situations as 'among them, some refused to kill and others stopped killing.'[54]

Personality Effects on Behaviours

Research has shown that the power of the situation is more dominant than personality when it comes to peoples' behaviours in stressful situations. However, if the power of the situation was all powerful 35 percent of teachers in the Milgram shock experiment would not have stopped at 300 volts but would have gone on to the full 450 volts like the remainder. Therefore, personality plays a part in peoples' actions in stressful situations. Research has demonstrated that 'there is a variability of behaviour within situations: different people behave differently through they may find themselves in similar circumstances.'[55] Professor Jessica Wolfendale's work on war crimes adds to the idea around personality,

> Even if one accepts that stress, fatigue, and fear can play a role in the commission of war crimes, it is natural to think that whether a combatant commits war crimes will

---

50 Browning, *Ordinary Men*, 159.
51 Ibid., 185.
52 Ibid., 171.
53 Goldhagen, *Hitler's Willing Executioners*, 250.
54 Browning, *Ordinary Men*, 188.
55 Mastroianni, 'The Person-Situation Debate', 7.

The S-CALM Model

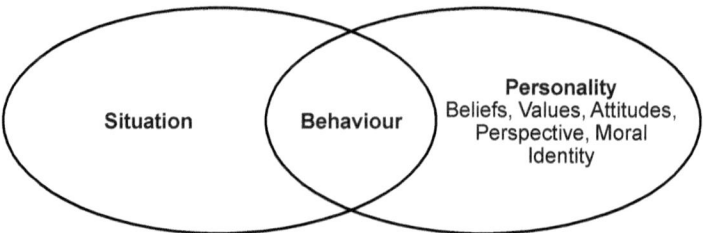

Figure 7.1 Situation and Personality

ultimately be determined by their ability to deal with those situational forces – by their strength of character, in other words. In this view, war crimes occur because military personnel lack self-control, empathy, leadership skills, and/or possess morally bad character traits such as sadism or cruelty.[56]

In Roth's, *Hearts of darkness: 'perpetrator history' and why there is no why*, there was further research around personality which produced an issue known as the 'Smile Problem'. During the research for the paper most people who committed abuse showed remorse, but 'for many who killed [they] did so uncoerced, indeed did so with enthusiasm and relish.'[57] The research demonstrated that personality does have some weight in ethical decision-making. This idea of how personality combines with the power of the situation to form behaviour is expressed in Figure 7.1.

In this respect personality has been termed as an individual's beliefs, values, attitudes, perspectives and their moral identity. For unethical actions it is the moral identity which is the most important aspect. Lawrence Kohlberg considered that people progress through six distinct stages of moral reasoning with the lowest level focused on obeying rules and self-interest. However, 'most adults operate a little higher on the scale, making decisions on the basis of normative influences, such as rules, regulations, and the opinions of others (most military personnel function at this level).'[58] There is then a small group, which should include military leaders who have had

---

[56] Wolfendale, *The Causes of War Crimes*, 275.
[57] Paul Roth, 'Hearts of Darkness: 'Perpetrator History' and why there is no why', *History of the Human Sciences*, Vol. 17 (2004), 214.
[58] Bradley, J. and Shaun Tymchuk, 'Assessing and Managing Ethical Risk in Defence', *Canadian Military Journal*, Vol. 13, No. 4 (2013), 13.

good ethical training and education, which operates at the higher levels and make their moral judgement based on ethical principles. For leaders this is important as a leader who is motivated by their moral identity is encouraged to act in an ethical way. It has been shown 'that the behaviors [sic] displayed by ethical leaders can "trickle down" to employees' therefore 'ethical leaders help develop group norms for how to treat others.'[59] Consequently, a leader who is operating at a higher stage of Kohlberg's moral reasoning scale and displays a good moral identity will influence their team to most likely act in an ethical manner in stressful situations.

Another piece of major research into how personality affects individuals' behaviours in unethical actions was conducted by King's College London. Their research investigated people who committed unethical acts and identified five different groups which they termed: 'perpetrators, bystanders, facilitators, enablers and heroes.'[60] Each group played a different part in an unethical action. These groups are summarised below:

- *Perpetrators.* Perpetrators are the zealots, the directors of violence. In a study of the Rwandan Genocide, Bhavnani believed that 'perpetrators were further split into two groups: Willing Executioners and Reluctant Perpetrators.'[61] The Willing Executioners were a small minority of participants and normally the ring leaders. They were able to compartmentalise their feelings and tended to be proud of their achievements. The Reluctant Perpetrators where the majority of participants who tended to 'go with the flow'. Nevertheless, they tended to feel discomfort and unease about their actions and suffered cognitive dissonance. They generally had to find a coping mechanism, often alcohol or drugs
- *Facilitators.* Facilitators take action that contributes to the perpetrators work
- *Enablers.* Enablers allowed the conditions to be set for the perpetrators and did not prevent their actions

---

59 David Mayer, Karl Aquino, Rebecca Greenbaum and Maribeth Kuenzi, 'Who Displays Ethical Leadership, And Why Does It Matter? An Examination of Antecedents And Consequences Of Ethical Leadership', *The Academy of Management Journal*, Vol. 55, No. 1 (2012), 154.
60 King's College London, Centre for Military Ethics, Armouring Against Atrocity, 2016, https://militaryethics.uk/en/course/library.
61 R. Bhavnani, 'Ethnic Norms and Interethnic Violence: Accounting for Mass Participation in the Rwandan Genocide', *Journal of Peace Research*, Vol. 43 (2006), 651.

- *Bystanders.* Bystanders tended to know what was happening but took no action to stop it
- *Heroes.* Heroes took positive action to stop perpetrators. One of the most well-known heroes was Warrant Officer Hugh Thompson, whose actions, as explained earlier, stopped the My Lai massacre.

Some in the military are in the perpetrators, facilitators, enablers or bystanders space and either commit, approve or most likely take no action when abuse is conducted. This abuse may not be the big atrocities that have been used as case studies in this book but might be low level unacceptable behaviour. This unwillingness to take action is not new, the Roman writer Tacitus commented that 'the worst crime was dared by a few, willed by more and tolerated by all.'[62] There is a vital part that ethical education has to play in changing this state, when soldiers are educated and trained in an ethical toolkit, such as the S-CALM model, they are more likely to climb into Kohlberg's higher stages and stop being bystanders to unethical actions and become what King's College London call heroes.

Personality Case Study – Abu Gharib Abuse

A well documented example of how personality can affect individuals' reactions to the power of the situation is in the well documented case of the Abu Ghraib abuse, which this book has visited a few times already. In this case study the central actors in the abuse can be divided into the categories identified by King's College London as follows:

- *Perpetrators.* The main perpetrators in this case were Charles Garner and Lynndie England. The activity carried out the abuse of the detainees. In Bhavnani's terms they were willing executioners
- *Facilitators.* The main facilitator in this case was Staff Sergeant Fredrick. As leader of the night shift he allowed Garner and England to conduct their abuse without censure. To make matter worse, he also occasionally joined in with abuse
- *Enablers.* The main enablers for the abuse were Brigadier General Janis Kaplinski and Lieutenant Colonel Jerry Phillabaum. Brigadier General Kaplinski was the brigade commander of 800th Military

---

[62] Good Reads. https://www.goodreads.com/quotes/9429606-the-worst-crimes-were-dared-by-a-few-willed-by.

Police Brigade and Lieutenant Colonel Phillabaum was commanding officer of 320th Military Police Battalion. These officers did not train their soldiers, including Garner and England, in International Humanitarian Law. They also failed to supervise Staff Sergeant Fredrick and the night shift at Abu Gharib. Thus, they set the conditions for the abuse to take place
- *Bystanders.* There were many members of 372nd Military Police Company and civilian contractors working in Abu Gharib that were aware of the abuse being conducted but did not have the level of moral courage that allowed them to either intervene or report what was occurring
- *Heroes.* Sergeant Joe Darby was the hero in this case. He considered the abuse to be against the moral code of the US Army and reported the actions of Garner and England.

This case study demonstrates that unethical behaviour is not just committed by the perpetrators but is allowed to take place by others not acting. A key aim of a leader is to ensure that their teams do not contain perpetrators and to try and educate their team to become heroes.

8

S-CALM:
C – Common Behaviour (Group)

Introduction

I was Battalion Second in Command in the UK, and the unit had a few weeks gap between operational commitments. The Commanding Officer was keen that we practice some conventional warfighting training. The companies organised low level tactical exercises and I organised a series of command post exercises to practice our battalion headquarters procedures. These involved conducting battle group planning and then giving battle group orders. After a week of planning a series of missions, we thought we had regained some of the skills lost during our counter insurgency deployments and decided to invite the platoon commanders to come and watch a battle group orders for their education. The plan was for an offensive action and orders to be given to the company commanders in front of the gathered platoon commanders who sat at the back to observe. When the formalities ended, I asked the platoon commanders if they had any questions on how the battle group orders had been given or on the plan or process. A second lieutenant, fresh from the Royal Military Academy Sandhurst, raised a hand and made the following comment 'Sir, thank you for the opportunity to observe the orders, I noticed that you had not allocated a reserve for the operation'. The challenge was correct, neither the planning team nor the company commanders had spotted this mistake. The mistake was due to our investment in the plan and as a team we had suffered groupthink. It took an independent observer who was not emotionally invested in the plan to point out the obvious mistake. Therefore, it was not the platoon commanders who had been educated that day, but us. In future planning, we did not use the Intelligence Officer to red team our planning as was considered the norm, but always appointed an independent devil's advocate to observe the plan and spot these mistakes.

In this chapter each of the six common behaviours that affect a team will be studied with the aim of recognising what the behaviour might look like in a stressful situation. The main mitigation of common group behaviours in the S-CALM model is achieved through the application of accountability, leadership and the moral compass, which will all be explored in later chapters. Nevertheless, where appropriate some mitigating ideas will be suggested. There will also be a positive case study example for each of the behaviours demonstrating a situation when an individual or team overcame a behaviour.

Groupthink

As discussed in chapter 4, groupthink is one of the more prevalent common behaviours identified within militaries. Leaders and team members should be scanning activities and decision-making to identify main groupthink symptoms. As a reminder these are: a feeling of invulnerability, group morality, group pressure, self-censorship and the appointment of mindguards. One of the best defences against groupthink within a team is the appointment of a devil's advocate or what the British Army calls a 'red team'. The British doctrine, *Joint Defence Publication 04*, explains how 'a red team can monitor groupthink bias and ensure that sufficient analysis and debate has been completed before a policy, plan or strategy is decided.'[1] Those appointed as a red team must have the standing within the team to be able to challenge the collective thinking of the group. If possible, it is useful if the red team is independent of the actual group, which means they are not exposed to peer pressure and the other common group behaviours discussed. This allows them to challenge ideas that the group, due to the symptoms of groupthink, might believe unquestionable. On this point, British doctrine states that 'overcoming groupthink requires acceptance of authentic dissent; commanders should be aware of this and both encourage and acknowledge dissenting views.'[2] Once a decision or action has been challenged by the red team, a leader needs to know what to do next. Irving Janis, who first

---

1 MOD, *Joint Doctrine Publication 04*, 52.
2 Ibid., 44.

The S-CALM Model

identified groupthink, believed that if a decision was challenged 'second chance meetings should be held to reconsider the decision once it has been reached and before it is made public.'[3] The key point is that a system of ensuring that teams do not fall into the symptoms of groupthink must be put in place. It is also important that each member of a team take responsibility for their actions and are a critical evaluator of the team's collective action. Therefore, the leader must encourage an open climate of giving and accepting criticism and be open to challenge. More details on giving and receiving challenges are in chapter 10.

Groupthink Case Study – Trooper Ben Griffin in Iraq

In early 2005, G Squadron, 22nd Special Air Service Regiment was on counter insurgency operations in Baghdad, Iraq. The SAS was working within a multi-national SOF (Special Operations Force) operation under US leadership. One of the soldiers in G Squadron was Trooper Ben Griffin and after three months in Baghdad, he 'told his commander that he was no longer prepared to fight alongside American forces.'[4] He said the reason for this was that he believed he had seen illegal acts by US troops, claiming they viewed all Iraqis as "untermenschen" - the Nazi term for races regarded as sub-human.' He also 'alleged that American forces were routinely using too much force and that these were illegal tactics.' He concluded that 'I did not join the British Army to conduct American foreign policy.'[5] In consequence, he applied and was granted permission to resign from the British Army. 'He expected to be labelled a coward and to face a court martial and imprisonment.' However, he was treated well by the MOD and Griffin's testimonial described him 'as a "balanced, honest, loyal and determined individual" who has "the courage of his convictions".'[6] There had been a growing feeling of unease within the British unit that rules of engagement were not being correctly followed and that the force was in danger of being turned into a secret police force under US authority. Groupthink was strong within the SOF community but Trooper Griffin was able to overcome the group pressure, self-censorship and mind-guards to speak the truth about what was going on and to have the confidence to refuse to be part of it.

---

3 I. Janis, *Groupthink*. (Boston: Houghton Mifflin, 1982), 262.
4 Sean Rayment, 'SAS soldier refuses to fight in Iraq', *The Age*, 13 March 2006.
5 Ibid.
6 Ibid.

Risky Shift

Operations in the armed forces is a risk-taking business, so it should not be a surprise that risky shift is one of the most commonly used behaviours. As a reminder risky shift can be defined as the 'tendency for group discussion to produce group decisions that are more risky than the mean of members pre-discussion opinions, but only if the pre-discussion mean already favoured risk.'[7] The final part of this definition is important as it points to the fact that teams can be more risk adverse if they have overcome groupthink and discussed less risky options. In the original research, Stoner noted that 'the group decisions tended to be more cautious on items for which widely held values favored [sic] the cautious alternative and on which subjects considered themselves relatively cautious.'[8] Therefore in ethical decisions a team can make more cautious decisions if they are steered in that direction. The important point for a leader is to know when to step in and stop teams from pushing their boundaries and becoming less risk adverse, which can result in the spiral of violence. When there is time to consider risk taking before entering a stressful situation, a well-known and useful tool is the 4Ts. The 4Ts are a consideration of options that a leader has when they have calculated a risk and are as follows:

- *Tolerate.* When a risk is tolerated, it has been identified and accepted as an appropriate risk for the task being conducted. It is important to differentiate between a risk, which is calculated and a gamble which is not
- *Treat.* When a risk is to be treated, it has been calculated and a method for dealing with it has been identified
- *Transfer.* When a risk is to be transferred, it has been calculated and a method for dealing with it is not within the scope of the immediate team, but the wider organisation has a solution and it will be passed on to be handled
- *Terminate.* When the risk is considered too large for the task, the activity is terminated.

There will not always be time to conduct the 4Ts when faced with risky shift in a stressful situation. However, leaders should always aim to be

---

7 Hogg and Vaughan, *Social Psychology*, 663.
8 Stoner, Risky and Cautious Shifts in Group Decisions.

exemplars, set the moral tone and present a clear ethical vision, which will be studied in more detail in chapter 10.

Risky Shift Case Study – Nisour Massacre

The Nisour massacre, occurred when a team from the Blackwater private military company killed 17 Iraqi civilians and was outlined in chapter 4. As was identified earlier, the shooting started by one contractor in Nisour square quickly escalated until the whole team and helicopters overhead were firing into the square killing 17 civilians and wounding 20 others. Not all in the team opened fire, at the trial at least three members of the team said they did not open fire. Mr Frost, the commander of the second vehicle stated that he was 'upset because he had seen Iraqis shot although they posed no threats'; whilst the vehicle turret gunner, Mr Murphy, said that 'he never fired his weapon, but that he saw 'civilians shot and killed who were clearly no threat to anyone by his fellow Raven 23 members.'[9] Whilst Mr Mealy, the turret gunner in the first vehicle of the convoy said that he only 'identified four other Blackwater members who fired their weapons at Iraqis.'[10] Therefore, it appeared that the social norm was not for all to fire and the shooting was stopped as individuals began to notice that others in their team were not firing. An Iraqi witness recalled that at least one contractor was shouting no, no, no, and trying to get the others to stop shooting during the massacre. It emerged that there was only one contractor who continued to fire once the initial, confused shooting had taken place. Other team members told 'American investigators that during the operation at least one guard continued firing on civilians while colleagues urgently called for a cease-fire. At least one guard apparently also drew a weapon on a fellow guard who did not stop shooting.'[11] This was reinforced by an official statement which confirmed that the team '"had an on-site difference of opinion," he said. In the end, a Blackwater guard "got on another one about the situation and supposedly pointed a weapon,".'[12] Therefore, after the initial shooting, the team deescalated the situation and

---

9 '3 Blackwater Guards Called Baghdad Shootings Unjustified', New York Times, 16 January 2010, https://web.archive.org/web/20170227170643/http://www.nytimes.com/2010/01/17/world/middleeast/17blackwater.html.
10 Ibid.
11 'Blackwater Shooting Scene Was Chaotic', New York Times, 10 April 2009, https://web.archive.org/web/20090410140405/http://www.nytimes.com/2007/09/28/world/middleeast/28blackwater.html.
12 Ibid.

were even prepared to use violence against their own members to stop what many quickly realised was an unethical action.

Authority Bias

As noted in chapter 4, due to the hierarchical nature of military service obedience to an authority figure can be an issue that leads to unethical actions. Milgram tended to combine these two concepts, but because of this unique operating environment in the military it is vital that leaders explore authority bias separately. Robert Cialdini's *Influence: The Psychology of Persuasion*, identified two important aspects of authority bias. The first is clothing about which he states it can 'trigger our mechanical compliance.'[13] In the military uniform and rank can be seen and is especially strong when those in authority have different uniform and badges from their followers. The second is the use of titles, which represent authority as 'to earn one normally takes years of work and achievement' and there is a natural assumption of power.[14] The British Army recognises these points and advises that 'the abuse of authority to intimidate or victimise others, or to give unlawful punishments and orders, is illegal and unacceptable.'[15] A leader needs to be careful how they use their authority gained by position and reinforced by dress and titles. In many situations a team will take the word of an authority figure as orders and act on it even if the leader had not intended it to be an executive order.

In the RMAS 2023 research survey, a question was asked to establish the trainees' awareness of authority bias. The question asked trainees to agree with the following statement: You should always obey an order by a senior commander? For the cohort with no previous military training, the research noted that 'there was almost a reversal of responses here between the first and final timepoints as responses shifted gradually across the three surveys' with the majority of them understanding authority bias by the end of their training.[16] For research question two, which was completed by all trainees in their final term, the preponderance of British respondents recognised the need for intelligence disobedience, but 23 per cent agreed that all orders should be obeyed. The results for the international cohort were less clear as

---

13 Cialdini, *Influence: The Psychology of Persuasion*, 226.
14 Ibid., 222.
15 MOD, *Values and Standards*, 25.
16 Vincent and Muhl-Richardson, *An Analysis of the Effectiveness of the S-CALM Model of Ethical Leadership*, 16.

'47% agreed and 40% disagreed, suggesting differences in the perspectives of those making up the international cohort.'[17] This research demonstrated that even with training on intelligent disobedience it can be difficult for leaders to be aware of authority bias. Therefore, due to their position of power and influence in difficult situations leaders should remain impartial and refrain from stating their personal preferences before their team, who are likely to follow their leaders' opinions. As has already been stated the use of participative leadership, openness to challenge and an awareness of intelligence disobedience are useful means for reducing authority bias. More ideas on overcoming authority bias are explored in chapter 9.

Authority Bias Case Study – Admiral Sandy Woodward in the Falklands War

In 1982, Admiral Sandy Woodward commanded the Aircraft Carrier Group in the Falklands War. At Midday on 21 April 1982, an unidentified aircraft was detected at long range on route to the Carrier Group. Admiral Woodward ordered that a Sea Harrier be launched and it intercepted an Argentinian Air Force Boeing 707, which they believe was not armed. It changed course when it was intercepted and was nicknamed the Burglar by the Task Force Operation Centre. At 0230 hrs on 22 April, another contact was made 65 miles from the Carrier Group, again Sea Harriers where dispatched and the Burglar quickly headed for Argentina. That night at 2000 hrs the Burglar again made for the Carrier Group. Admiral Woodward recalled that 'by now, everyone was in a high state of agitation.'[18] Therefore he asked permission to shoot this aircraft down from Fleet Headquarters in the UK and this was agreed providing that 'a) he came within a certain specific range limit and b) we had positive identification that he was indeed, the Burglar.'[19] At 1134 hrs the next day the Burglar returned and Sea Harriers where deployed to intercept, but the aircraft vanished before it was reached. Admiral Woodward commented that the Burglar was 'becoming something of a habit, a bad one and unwelcome with it.'[20] Just after sunset the Burglar appeared again and was heading directly for the Task Force and in two minutes it would be within missile range, it was now too late to dispatch Sea Harriers to intercept it. Admiral Woodward wrote that 'the Burglar has been visiting us regularly

---

17 Ibid.
18 Sandy Woodard, *One Hundred Days: The Memoirs of the Falklands Battle Group Commander.* (London: Harper Collins, 2003), 143.
19 Ibid.
20 Ibid., 144.

now for three days. It is time to remove him.'[21] Admiral Woodward asked if there were any scheduled commercial flights in the area and he was told no. With one minute to go, everyone was ready for Admiral Woodward to give the authority to launch a missile, however he asked that the flight path of the Burglar be very quickly plotted. When this was quickly done, the plane's flight path if extended each way was a direct line from Durban in South Africa to Rio de Janeiro in Brazil. Admiral Woodward immediately ordered 'weapons tight', to all ships, which restricted anyone firing at the aircraft. A Sea Harrier was dispatched and it identified the aircraft as a Brazilian airliner.

If Admiral Woodard had given the authority for missile launch the airliner would have been immediately shot down. There rightly would have been world condemnation, as happened just after on 1 September 1983 when the Soviet Union shot down a Korean airliner near Japan thinking it was a spy plane killing 269 people. There is little doubt that the US would have removed support to the UK and most likely that the Task Force would have been recalled before it had even got close to the Falklands Islands. The First Sea Lord, Admiral Sir George Zambellas, said of Woodward that he had displayed 'inspirational leadership and tactical acumen ... [which] was a major factor in shaping the success of the British forces in the South Atlantic.'[22] Admiral Woodward was under considerable pressure to make a vital decision with less than one minute to go, had he given the order it is extremely unlikely that anyone in the Carrier Group would have questioned his authority with disastrous results.

Othering

As discussed in chapter 4, militaries often use othering when in combat. To understand opinions on dehumanisation and othering, the RMAS 2023 research survey asked the question: It is acceptable to use nicknames for the enemy that you are fighting? The results for the cohort with no previous military training were consistent in the first and final survey with '26% of

---

21 Ibid.
22 'Falklands War Admiral Sandy Woodward dies aged 81', BBC News, 5 August 2013, https://www.bbc.co.uk/news/uk-23575534.

respondents believed that using nicknames for the enemy was acceptable.'[23] It was interesting that the result for the second survey, immediately after S-CALM model education, was considerably lower at 15 per cent, but went up again possibly due to absorbing othering as the trainees developed a stronger subunit ethos in training. For research question two for all trainees, 66 per cent of the British cohort and 60 per cent of the international cohort disagreed that it was acceptable to use nicknames for the enemy. The research noted that '40% of international Ocdts [Officer Cadets] agreed with this statement (compared to 26% of UK Ocdts) suggesting that this was a much more binary issue for international Ocdts.' It finally concluded that 'it is possible that Ocdts are absorbing othering as they develop a stronger subunit ethos.'[24] Militaries use othering on a routine bases to build ethos and team *esprit de corps* and develop feelings of pride and loyalty within teams. It is very often done by developing unit nicknames, slang and badges all of which set the team apart from others. The building of a unit ethos is a traditional way of developing a unit's fighting spirit. However, General Sir Alan Cunningham delivering a speech called *Leadership* in 1944 accepted that unit traditions and ethos were important but that 'it had been proved throughout the war that new units who were well led would provide an esprit de corps which was second to none. It would not do to be carried away by tradition.'[25] As a leader it is important to build a unit's cohesion, without being negative and giving derogatory names to 'out-groups' such as other organisations, especially the enemy. The Australian Defence Force gives clear direction to its leaders on this point and their doctrine states that 'they' are never less than 'us'.'[26] Leaders must be quick to stop any othering that they identify as it can quickly deteriorate into dehumanisation and demonisation.

Othering Case Study – Lance Corporal Kylie Watson in Afghanistan

As the NATO conflict in Afghanistan went on, the Taliban developed a tactic that they had used against the Soviet Union of infiltrating their fighters into the governments security forces and conducting attacks from within, these were termed 'green on blue' attacks. The 'green' being Afghan government

---

23 Vincent and Muhl-Richardson, *An Analysis of the Effectiveness of the S-CALM Model of Ethical Leadership*, 16.
24 Ibid.
25 Dennis Vincent, 'Cunningham: On Leadership', *The British Army Review 180, Spring/Summer 2021* (2021), 110.
26 ADF, *Military Ethics*, 24.

security forces and the 'blue' being NATO forces. A report said that 'concern about the issue among US and NATO officials has now become so great that trust between them and Afghan security and military officials is at an all-time low.' It went on to report that 'more than 70 NATO troops have been killed by Afghan colleagues in recent years, leading to what some diplomats say is an irretrievable breakdown of trust between the two sides.'[27] The 1st Battalion, the Duke of Lancaster's Regiment, deployed to Helmand Province into this arena of mistrust in 2010. Lance Corporal Kylie Watson was the Royal Army Medical Corps medic for 9 Platoon, C Company, of the 1st Battalion, the Duke of Lancaster's and was based at Checkpoint Azadie.

Lance Corporal Watson was on a patrol alongside the Afghan security forces when an Afghan soldier was shot. An account recalled how 'Watson gave medical care in exposed open ground for 20 minutes. She had previously run 100 metres in full view of the enemy under sustained fire to give lifesaving first aid to an Afghan soldier who had been shot twice in the pelvis.'[28] In addition to dealing with the wounds Lance Corporal Watson had to deal with the wounded Afghan comrades who tried to prevent the treatment as they did not want the soldier treated by a woman. Lance Corporal Watson recalled that 'we were under heavy fire and whenever you've got a casualty and it's your job to treat them you just go and treat them - I didn't really think about it much at the time.'[29] For this action she was awarded the Military Cross 'her citation read: "Watson's immense courage, willingness to put her own life at risk and absolute bravery saved the life of one warrior and acted as an inspiration to her platoon and their Afghan National Army partners."'[30] In 2011, there were 21 NATO soldiers killed in 'green on blue' attacks and there was a degree of mistrust between NATO and Afghan troops with a certain degree of othering to be found. Regardless of these strained relationships, Lance Corporal Watson risked her life to save the life of a wounded Afghan solider.

---

27 'Nato's crisis of trust in Afghanistan', BBC News, 9 March 2012, https://www.bbc.co.uk/news/world-asia-17219153.
28 'Military Cross winner Kylie Watson 'shocked by award', BBC News, 28 March 2011, https://www.bbc.co.uk/news/uk-northern-ireland-12886147.
29 Ibid.
30 'Army medic Kylie Watson awarded Military Cross', BBC News, 27 March 2011, https://www.bbc.co.uk/news/uk-12873091.

Dehumanisation

Dehumanisation is one of the two behaviours, with demonisation, that are not routinely used by armed forces members on a day-to-day basis. Leaders need to be alert for the display of the two types of dehumanisation: animalistic and mechanistic. Animalistic dehumanisation is displayed by shows of emotion towards others such as contempt and disgust; whilst with mechanistic dehumanisation people are seen and treated as objects. People must always be seen as humans and the value of life must be emphasised. A leader needs to display respect for their teams and encourage respect for others ensuring that all are treated with dignity, equality and worth. When on operations or in combat, leaders must be attuned to the terms used for 'the enemy' and any euphemisms for them and their unethical treatment must be stopped before the spiral of violence can begin. These ethical principles will be expanded more in chapter 11 which will explore the need for a sound moral compass.

Dehumanisation Case Study – Christmas Day Truce 1914

When the First World War started in 1914, Great Britain and Germany were on close terms. They shared a royal family and had been allies against the French for hundreds of years. In 1914 there were 'around 53,000 Germans resided in the UK, making up the third largest minority group after the Irish and Jewish.' There were around 27,000 Germans in London and many of them were 'barbers, bakers or restaurateurs (Soho's Charlotte Street, was nicknamed Charlottenstrasse because of the number of German restaurants and bakeries).'[31] The British Government set about a campaign to dehumanise the Germans with the Prime Minister Herbert Asquith giving a speech in October 1914 accusing the Germans of 'devastation and destruction worthy of the blackest annals of the history of barbarism.'[32] There was also propaganda including posters depicting the Germans as beasts and monsters. It was not long before 'the British acquired a talent for hating Germans' but it should be noted that, 'the Germans excelled themselves in hating the British – or, more specifically, England.'[33] For in Germany a similar propaganda campaign was being conducted to dehumanise the British and all over Germany could

---

31 'Thinkers or Junkers? Germans in England 1860-1920 & Beyond', Anne Hill Fernie, 2018, https://raggeduniversity.co.uk/2018/09/15/germans-in-england-1860-1920/.
32 Malcolm Brown and Shirley Seaton, *Christmas Truce*. (London: Pan, 1994), 4.
33 Ibid., 5.

be heard 'HATE, HATE, the accursed ENGLISH HATE.'[34] These campaigns continued into the winter of 1914 and by Christmas time it was considered that the two armies had dehumanised the other side and were expecting the other to take advantage of this period when their enemy might not be as vigilant as normal. The British High Command issued an order on Christmas Eve 1914 that 'it is thought possible that the enemy may be contemplating an attack during the Xmas or New Year. Special vigilance will be maintained during this period.'[35] The German Command felt the same, Private Hugo Klemm the 133rd Saxon Infantry Regiment recalled that the unit commander 'emphasised that for that day and the following days special alertness' was required.[36]

However, over the Christmas period is it estimated that around 100,000 British and German troops were involved in the informal cessations of hostility along the Western Front. It was noted that the 'propaganda had far outstripped reality and when the enemies met face to face they found that they were not only human but also on the whole, likeable.'[37] Many of the German troops had been waiters or barbers in London and returned to Germany at the start of the war and been called into the reserves. A '1/Leicester Tommy remembers a voice saying in clear English, 'Hello there, hello there, we are Saxons, you are Anglo-Saxons. If you don't fire, we won't fire.'[38] Soldiers from both sides were soon meeting between the trenches and exchanging gifts such as food, tobacco, alcohol, and souvenirs like buttons and hats. They also took the opportunity of burying the dead and one officer noted that 'it was a curious situation – enemies working together to sort out and bury the men they had jointly killed.'[39] Another young officer, Lieutenant Malcolm Kennedy recalled that 'for the time being, all horrors and discomforts of the War seemed to be forgotten.'[40] The writer Bernard Brookes reported that 'the Germans have no bitter feelings towards us' and the painter Bruce Bairnsfather stated that 'there was not an atom of hate on either side that day.'[41] Despite all the propaganda the troops had managed to establish a human connection with those they were fighting for this short period and treated each other with dignity and respect.

---

34 Ibid., 7.
35 Ibid., 55.
36 Ibid.
37 Ibid., 94.
38 Ibid., 82.
39 Ibid., 84.
40 Ibid., 54.
41 Ibid., 93.

Demonisation

As previously mentioned, demonisation is not a behaviour that a leader would expect their team to display routinely, but in volatile, uncertain, complex and ambiguous situations it is inevitable that there is a tendency to try to simplify the conflict to a black and white situation. This results in the solution that we are good and the enemy is bad. But as previously explained, combat is very rarely that simple and leaders should avoid presenting the enemy as evil. A leader needs to present a balanced understanding of their adversaries and communicate a sense of empathy with them. Some techniques for the leader are outlined in the focus on leadership in chapter 10.

Demonisation Case Study – First Lieutenant Josef Sibille in Belarus 1941

The German Army's 691st Infantry Regiment conducted occupation duties of Belarus during Operation Barbarossa in 1941. The unit had only recently arrived in Belarus, having come from occupation duties in France. The 1st Battalion was deployed in the area around Krucha. On 7 October 1941, the 1st Battalion's commander, Major Alfred Commichau, issued an order to the unit's company commanders directing them to kill all Jews in their respective areas. Lieutenant Hermann Kuhls, a 33 year old Nazi and a member of the SS, commanded the 2nd Company and he immediately ordered the company to begin shooting Jews. The 3rd Company was commanded by Captain Friedrich Nöll, who was a First World War veteran and known to be a little indecisive. He initially hesitated and then told senior NCOs to order the company to carry out the killings. However, 'the commander of 1st Company, 47-year-old First World War veteran and Nazi Party member First Lieutenant Josef Sibille, reportedly refused the order outright.'[42] Lieutenant Sibille stated that Major Commichau had said, '"as long as the Jews are not eliminated, we will not have any peace from the partisans. The Jewish action in your area must therefore be completed in the end." Sibille related that this order caused "anxious hours and a sleepless night" until he made his decision.'[43] After repeated urgent calls from the battalion commander, Lieutenant Sibille informed Major Commichau 'that "my Company would not shoot any Jews, unless we catch the Jew with the opposing partisans". He explained that he

---

[42] Jody Prescott, et al. *Ordinary Soldiers: A Study in Ethics, Law and Leadership*. West Point: Center for Holocaust and Genocide Studies at West Point, 2014, 9.
[43] Ibid., 15.

could not "expect decent German soldiers to dirty their hands with such things".'[44] Major Commichau then asked 'Sibille when he would "be hard for once" to which Sibille replied, "in this case, never".'[45] Major Commichau gave Sibille three days to carry out the order. Lieutenant Sibille did not carry out the order to kill the Jews, not only for the professional reason given to Major Commichau but also on 'a deeper moral objection.'[46] Apart from being accused of being soft by other officers in the Battalion, 'Sibille apparently suffered no further repercussions from his refusal to obey the execution order.'[47] It appears that he was just one of the '85 documented cases in which the Wehrmacht, SS, or police service members refused to shoot jews.'[48] Lieutenant Sibille provides an excellent example of a leader who refused to demonise an 'out-group' which would have made it easier to dehumanise and ultimately abuse and murder them.

Summary

The last two chapters have highlighted how a leader might recognise and mitigate the 12 cognitive behaviours that have been identified as common when the situational influencers multiple the effects of a stressful military situation. The positive case studies have been used to highlight experiences where leaders have faced difficult situations but have been able to overcome the behaviours that could have resulted in unethical action. It is easy to look at these cases and not draw lessons for our own decision-making as 'everyone can see the collective mistake, but no one draws any individual conclusions.'[49] However, the reason for studying these common behaviours is that research has ascertained that 'highly intelligent people will display fewer reasoning biases when you tell them what the bias is and what they need to do to avoid it.'[50] An important part of moderating the effects of these behaviours is to try and ascertain which ones an individual uses in routine circumstances. It is likely that these behaviours, if used in normal day to day business, will become a default behaviour when under stress.

---

44 Ibid.
45 Ibid.
46 Ibid.
47 Ibid., 16.
48 Ibid.
49 Oliver Sibony, *You're About To Make A Terrible Mistake: How Biases Distort Decision-Making and What You Can Do to Fight them.* (Croydon: Swift Press, 2021), 163.
50 Keith Stanovich and Richard West, 'On the Relative Independence of Thinking Biases and Cognitive Ability', *Journal of Personality and Social Psychology*, Vol. 94, No.4 (2008), 690.

The S-CALM Model

Common individual Behaviours	Common Group behaviours
Social Comparison Theory/ Conformity	Group Think
De-individuation	Risky Shift
Obedience	Authority Bias
Cognitive Dissonance Theory	Othering
Bystander Effect	Dehumanisation
Status Quo Bias	Demonisation

Table 8.1 Common Individual and Group Behaviours

There is now an opportunity to conduct some self-reflection and consider what individual and group behaviours you and your immediate team or group use every day. Below there is room for you to note down the two most dominant behaviours in each category, but you might want to note more if appropriate. As a reminder the common individual and group behaviours they are at Table 8.1.

Identify the two dominant individual behaviours that you routinely use?

- _____

- _____

Identify the two dominant group behaviours that your team routinely use?

- _____

- _____

Once you have completed your common behaviours, I recommend that you find a trusted peer, who knows you well and with whom you feel that you have psychological safety. With this peer confirm that they agree with your selection of individual behaviour choices for yourself and your group behaviour choices for your team. This should enable you to confirm what your most likely default behaviours will be in a stressful situation. This in turn should enable you to firstly look for their signs and importantly have considered how you intend to mitigate their effects on your actions. Having got to this stage, the next 3 chapters will investigate how using accountability, leadership and the moral compass can enhance our moderation of these behaviours.

9

S-CALM:
A – Accountability

Introduction

I was Battalion Second in Command of the third unit into Kabul, Afghanistan in 2002 following the fall of the Taliban. We were based in Camp Souter and received a series of official briefs on the situation from the unit that we took over from. During these briefs, the unit continually referred to the Afghanis as 'rag heads'. This was a derogatory term for the locals they had picked up from the US Army. At that time Kabul was packed with internally displaced people who had fled the fighting against the Taliban and were living in huge, tented encampments on the plain around Kabul. They had left their homes with nothing but their dignity. The commander and I discussed the situation and concluded that if we took away the dignity of these people who had nothing by calling them names and did not show them respect, there was no reason why they would support us. I therefore warned the battalion that it was not acceptable to disrespect the locals by calling them derogatory names despite what they had heard. I also told them that we would discipline the first solider caught doing so by sending them back to the UK. We duly caught someone and sent them back to the battalion's base in Chepstow. This sent a message to the remainder of the battalion that we would not accept othering of the Afghanis who were to be treated with dignity. Once the battalion knew that they would be held accountable for their actions, the calling of names ceased and the Afghanis were treated with the proper level of respect for the remainder of the tour of Kabul.

Military personnel understand that they are accountable for their own actions. However, the ethical leader is also accountable for the training and preparation of their teams and their subsequent actions. Accountability can be defined as 'holding individuals legally responsible for their behaviour and imposing some form of sanction on them if warranted.'[1] Another definition of accountability is the 'control and direct administrative behaviour by requiring answerability to some external authority.'[2] This idea of an outside power is repeated in a third definition which states that 'ethical behavior, [sic] in short, required the presence of external accountability mechanisms in all their various forms.'[3] Later in this chapter the four common external influences will be examined, however first the approaches of the British, US and Australian militaries to accountability will be explored.

Accountability in the Military

In British Army doctrine, the actual term accountability is not used, but it states that 'leaders must accept responsibility for themselves and their people … leaders encourage those they lead to accept responsibility for their own behaviour.'[4] Although this mirrors the points made above, responsibility is not synonymous with accountability. An appointed commander has responsibility for themselves and their teams, but accountability goes beyond the mere action of being responsible. US Army doctrine states that 'leaders take responsibility for their actions and those of their subordinates' and 'leaders should not intentionally issue vague or ambiguous orders or instructions to avoid responsibility.'[5] Again this is good practice but avoids the consequences of those who order. US Army Special Forces are more direct and state that 'accountability means you are always responsible for your choices, even if you are doing what someone else has asked of you.'[6] A study of 271 senior US Army leaders, the people who give those orders found that the study's 'participants emphatically believe that personal and professional accountability is a critical vehicle to deal with the temptations of command.'[7]

---

1 Susan Kemp, *British Justice, War Crimes and Human Rights Violations* (Switzerland: Palgrave MacMillan, 2019), 2.
2 Dorothy Olshfski, 'Accountability and Ethics in the Abu Ghraib Scandal', Paper delivered to the Ethics and Integrity of Governance Conference, Leuven, Belgium, 2005.
3 Melvin Dubnick, 'Accountability And Ethics: Reconsidering the Relationships', *International Journal of Organization Theory and Behavior*, Vol. 6, No. 3 (2003), 406.
4 MOD, *Army Leadership Doctrine*, 2-10.
5 US Army, *ADP 6-22*, 2-2 and 2-7.
6 US Army, *A Special Operations Forces Ethics Field Guid*, 42.
7 Clinton Longenecker and James W. Shufelt, 'Conquering the Ethical Temptations of Command: Lessons from the Field Grades', *JPME Today*, 2nd Quarter 2021 (2021), 40.

The study went on the conclude that,

> Our leaders stated that creating an ethical leadership culture is a critically important guardrail for those in their command structure, as well as themselves. When senior leaders lead by example, operate with transparency, and help establish an ethical/moral command climate, employing all the tools available to them, they create not only downward accountability for their people but also upward accountability for themselves.[8]

This study demonstrated a good understanding of the role of ethical leaders to understand accountability. The Australian Defence Force doctrine reinforces this idea stating that 'all leaders are responsible not only for their own actions, but also for setting the ethical environment in their team.'[9] Their doctrine also underlines the point that 'commanders and leaders will be held to account for the ethical climate and for the behaviours of their teams.'[10]

However, accountability for ethical actions starts before the action takes place, as if a leader and their team enter a stressful situation 'without taking time to consider where they stand morally and ethically, it is already too late.'[11] Leaders must prepare their teams as best as they can for possible situations. The soundest way to do this is by education and training. The Australian Defence Force order that 'leaders at all levels have a responsibility to mentor and educate their personnel on the ethics of a particular conflict or situation.'[12] This education is important as research has shown that 'leaders who work in strong ethical contexts that support ethical conduct will better be prepared to handle morally intense situations and demonstrate their ethical leadership.'[13] Although these militaries understand the requirement for accountability, there is no clear explanation of exactly what it is and how to implement it.

---

8 Ibid., 42.
9 ADF, *Military Ethics*, 5.
10 Ibid., 37.
11 Mitchell Hall, 'Why Leaders need a Morality Check', *United States Naval Institute*, Vol. 132, No.4 (2006), 68.
12 ADF, *Military Ethics*, 5.
13 Brown and Trevino, 'Ethical Leadership: A Review and Future Directions', 599.

Type	Measure	Who
Hierarchical	Rules and Procedures	Subordinates to Superior
Professional	Responsibilities and Obligations	Expert to Leader
Legal	Laws and Conventions	All to Law Makers
Political	Stakeholders and National Interest	All – 'Strategic Corporal'

Table 9.1 Types of Accountability

Four Types of Accountability

Research has demonstrated that accountability can be broken down into four types in a military organisation, hierarchical, professional, legal and political. These four types, how they might be measured and who undertakes them are summarised in Table 9.1:

Hierarchical Accountability

Hierarchical accountability is the default for most militaries. It can be defined as being 'based on the ability of supervisors to reward or punish subordinates.'[14] Therefore in an orderly, tiered environment such as that which operates in the armed forces, it works via the chain of command with subordinates being accountable to their superiors. It tends to rely on the following of rules and procedures and has been described as 'bound by a set of values, discipline, tradition, unity and cohesion that form a professional military.'[15] It is dependent on personnel understanding the system and what they are empowered to do and what has to be authorised by their supervisor. The UK MOD's advice is that a leader must make sure that they 'know who is accountable for what; and that you bear personal responsibility for your contribution.'[16] The RMAS 2023 research survey asked trainees if: Accountability to your commander is the most important type of accountability? Research question one, for those that had no previous training was aimed at the effectiveness of training on the cohort over the year. The replies were split with around 40 per cent agreeing, 40 per cent

---

14 Romzek and Dubnick, 'Accountability in the Public Sector: Lessons from the Challenger Tragedy', 229.
15 Olshfski, 'Accountability and Ethics in the Abu Ghraib Scandal'.
16 MOD, *The Good Operation*, 15.

disagreeing and 20 per cent not sure. The research concluded that 'there was a wide distribution of responses across agree, unsure and disagree at all three timepoints, suggesting a high level of individual variation.'[17] The responses from all trainees was to establish their understanding of the military bias towards hierarchical accountability. The responses reflected this bias with:

> Over a third of UK Ocdts believe that hierarchical accountability was more important than legal, professional and political accountability. An equal weighting of these different forms of accountability would suggest that this figure should have been around 25%. There was a high 80% agreement by international Ocdts who believed that hierarchical accountability is the most important, which most likely aligns with the cultural requirement in many cultures to agree with the leadership.[18]

The research concluded that 'superficially, the Ocdts understand the need for legal accountability and to act within International Humanitarian Law' but that 'hierarchical accountability is still the most dominant form of responsibility, especially in the International Ocdt cohort.'[19]

One of the possible issues with hierarchical accountability is that leaders may believe that they need to conform to the environment. However, as explained earlier there can be no diffusion of responsibility or accountability when in command. It is therefore important for leaders to understand that 'accountability goes hand-in-glove with power and needs to be unpacked, called out, made explicit and transparent.'[20] There can be no diffusion of accountability by a leader who needs to be aware of the cognitive traps such as conformity and groupthink. Diffusion of accountability can lead to 'issues of misconduct in organisations, where problems that have been intrenched by history and tradition are inherited but not owned' by the leader.[21] There is also the danger of obedience with leaders following orders and 'control emanating from the higher levels of the hierarchy.'[22] There is also the opposite danger when units are dispersed or where there is a lack of supervision and

---

17 Vincent and Muhl-Richardson, *An Analysis of the Effectiveness of the S-CALM Model of Ethical Leadership*, 17.
18 Ibid.
19 Ibid., 21.
20 Olshfski, 'Accountability and Ethics in the Abu Ghraib Scandal'.
21 Crompvoets, *Blood Lust, Trust and Blame*, 5.
22 Olshfski, 'Accountability and Ethics in the Abu Ghraib Scandal'.

there is no extrinsic pressure to be accountable, in these situations military personnel can avoid being held accountable for their actions.

Professional Accountability

Professional accountability is a more intrinsic type of answerability where an individual benchmarks themselves against a high standard. It is 'dependant on the individuals understanding and identifying with the goals and ideals of the profession.'[23] In this type of accountability a subject matter expert is accountable to their leader for an action that the leader may not be able to check is correct but must trust the expert. Therefore, it is based on 'the central relationship [which] is similar to that found between a layperson and expert.'[24] In this situation subject matter experts understand their responsibilities and obligations and 'expect to be held fully accountable for their actions and insist that agency leaders trust them to do the best job.'[25] An example of professional accountability in the British Army would be the role of an aviation mechanic. Normally a corporal, they have the training to fix helicopters. They are normally in a Royal Electrical and Mechanical Engineer (REME) workshop and overseen by a junior officer. That officer may not have the training to do the various specialist roles the mechanics do, but when the mechanics confirm that the helicopter is fit to fly, the officer trusts their expert knowledge. The vital part of building professional accountability is both in the individual being well trained, having intrinsic motivation to do the best job possible and in mutual trust between those individuals and their leadership.

Legal Accountability

Legal accountability has been defined as being 'based on the relationship between a controlling party outside the agency and members of the organisation.'[26] It is concerned with abiding by laws and conventions and applies to all service personnel who are responsible to the law makers. For British personnel these laws and conventions include UK Law and International Humanitarian Law, known in the British armed forces as

---

23 Ibid.
24 Romzek and Dubnick, 'Accountability in the Public Sector: Lessons from the Challenger Tragedy', 229.
25 Ibid.
26 Ibid., 228.

the Laws of Armed Conflict and worldwide as the Geneva Conventions. British forces serving overseas are also answerable to local national laws. This accountability is important as it provides legitimacy when deployed. However, not all forces are equally answerable. Research in the US found that,

> In Iraq and Afghanistan of the 100 detainees who died in U.S. custody between 2002 and 2006, 45 are confirmed or suspected murder victims. Of these, eight are known to have been tortured to death. Only half of these eight cases resulted in punishment for U.S. service members, with five months in jail being the harshest punishment meted out.[27]

Upholding to legal accountability is vital for a military to be able to be identified as a legitimate force both at home and when deployed. When the law is broken, such as was explored in the Canadian Somalia affair, it can do a lot of damage to the legitimacy of the whole force.

Political Accountability

At the highest levels of military service political accountability seems simple, with senior officers responsible to stakeholders, such as politicians, and ensuring that national interests are protected. Therefore, in the British Army for example, the Chief of the General Staff is responsible for the Army and is accountable to the Secretary of State for Defence. Whilst this high-level political accountability may not seem applicable to most leaders at the junior level, in the age of the 'strategic corporal', actions conducted a low level can soon be of national importance. The concept of the 'strategic corporal' has been defined as an individual acting in such a way that they 'will potentially influence not only the immediate tactical situation, but the operational and strategic levels as well. His actions, therefore, will directly impact the outcome of the larger operation.'[28] Political accountability in the past was often not considered at the lower levels, but due to the power of social media and the speed at which what may appear minor actions in remote places can be shared, it is important that all have an understanding how their actions could be interpreted.

---

27 Peter Fromm, Douglas Pryer and Kevin Cutright, 'The Myths We Soldiers Tell Ourselves: and the Harm These Myths Do', *Military Review, September-October 2013* (2013), 63.
28 Charles Krulak, 'The Strategic Corporal: Leadership in the Three Block War: Operation Absolute Agility', Center for Army Lessons Learned, Fort Leavenworth (2002), 5.

British Army Standards

One of the major issues with the four types of accountability is that there is often friction between them. Research has shown that 'one or more of these accountability types may operate on the individual in a given situation and they may pull the decision-maker in different directions.'[29] Therefore the leader needs to find a way to keep this conflict of interest in balance. Generally, military leaders are aware of hierarchical accountability, due to the nature and structure of the military system, but the other three types are sometimes not considered. A British Army leader has a useful check list for ensuring their decisions are ethically accountable - the British Army Standards. The British Army Standards were initially defined in doctrine as 'the way in which we put our Values into practice, ensuring that everything we do is Appropriate, Lawful and Totally Professional.'[30] Since this initial inception, appropriate has been replaced by acceptable. They can be remembered by using the acronym LAP, for Lawful, Acceptable and Professional.

Lawful

When faced with a difficult situation the first question a leader can ask comes from the L of LAP, 'is this lawful'? As described earlier British military personnel are under the Armed Forces Act of 2006. This Act of Parliament reminds military personnel that 'all officers and soldiers are subject to the Criminal Law of England and Wales and are required to abide by it wherever they serve and at all times. All civilian criminal offences have been incorporated into Service Law' and 'Service Law creates additional offences.'[31] In addition to this domestic law, 'when deployed on operations, soldiers are subject to international law, including the laws of armed conflict, the prescribed rules of engagement and in some cases local civil law.'[32] Therefore leaders asking themselves 'is this lawful' will consider their knowledge of living in the UK combined with their training in the Law of Armed Conflict. The RMAS 2023 research survey asked a question to establish trainees understanding of the requirement to be lawful in their actions. The question asked them to agree with the statement that:

---

29 Olshfski, 'Accountability and Ethics in the Abu Ghraib Scandal'.
30 MOD, *The Army Leadership Code*, 10.
31 MOD, *Values and Standards*, 25.
32 MOD, *The Army Leadership Code*, 11.

You can break the rules of war if you believe that they are not useful for your mission? The responses from the cohort that had no previous military training was consistent with 98 per cent disagreeing and the research noted that 'their strong and consistent bias towards disagreeing with this question, indicating that Ocdts [Officer Cadets] had a good understanding of their legal duties and the requirement for legal accountability throughout.'[33] For Research question two, which explored the combined answers of all trainees, 99 per cent of British and 93 per cent of international trainees disagreed. The research concluded that 'it appears that there is a good understanding of the need to operate within the International Humanitarian Law.' Nevertheless, it went on to add 'there is some conflict with the responses to the Moral Compass questions below that need to be explored further.'[34] This conflict will be explored in chapter 11. The research demonstrated that although the trainees showed a bias towards hierarchical accountability, they were also fully aware of their legal accountability.

Acceptable

The second question a leader can ask in a difficult situation comes from the A of LAP, 'is this acceptable'? British Army doctrine states that acceptable behaviour 'is that which fosters team spirit and cohesion' and 'it is the duty of every member of the British Army to exhibit and promote acceptable behaviour, at all times and in all contexts.'[35] The doctrine also highlights that 'unacceptable behaviour undermines trust and cohesion, directly impacting operational effectiveness.'[36] The need for operational effectiveness underpins the requirement for acceptable behaviour both in barracks on a day to day basis and on operations. A way to measure unacceptable behaviour is known as the Service Test. The Service Test asks the question 'Have the actions or behaviour of an individual adversely impacted or are they likely to impact on the efficiency or operational effectiveness of the Service?'[37] If the answer to the question is yes, then a leader will need to act, which could be administrative or disciplinary action. The decision on what action is to be taken 'will depend on the circumstances of each case measured against the

---

[33] Vincent and Muhl-Richardson, *An Analysis of the Effectiveness of the S-CALM Model of Ethical Leadership*, 17.
[34] Ibid.
[35] MOD, *Values and Standards*, 26.
[36] MOD, *The Army Leadership Code*, 10.
[37] MOD, *Values and Standards*, 30.

Service Test.'³⁸ The Service Test sets the standard but is generally used in non-operational circumstances. On operations the leader must set the standard of acceptable ethical behaviour for their teams and ensure that they know they will be held accountable for breaking this standard. This requires a degree of transactional leadership as they reward those that behave well and discipline those who do not.

Professional

The final question a leader can ask in a difficult situation comes from the P of LAP, 'is this professional'? Professional behaviour is defined in British Army doctrine as 'how you act or react, intentionally or unintentionally and whether on or off duty.'³⁹ A more detailed definition states that,

> It should go without saying that all ranks must always conduct themselves in a manner that is totally professional. Firstly, this approach is critical to maintaining operational effectiveness and achieving the mission we are set. Secondly, it is necessary to protect and promote the Army's proud reputation, which has been hard won by endeavour and sacrifice.⁴⁰

Most governments invest a lot of time and money into training their armed forces to be accomplished in the profession of arms. Sir James Glover's research concluded that there was an important need for professionalism as 'it breeds, or should breed, the thinking soldier.'⁴¹ The thinking solider is able to recognise when their actions are not to the right standard and correct them. Dahr Jamail in *Will to Resist: The Soldiers Who Refuse to Fight in Iraq and Afghanistan,* relates a story from Baghdad in 2007 of how an infantry platoon which had taken serious losses on an 11-month tour knew that they should not go on patrol. They believed that the attrition had made them no longer able to 'function professionally' and were 'concerned that their anger could touch off a massacre of Iraqi civilians.'⁴² Although a difficult situation to admit, this demonstrates a mature attitude to the profession of arms and the

---

38 Ibid.
39 Ibid., 28.
40 MOD, *The Army Leadership Code,* 11.
41 James Glover, 'A Solider and His Conscience', *Parameters,* Vol. 13, No. 1 (1983), 56
42 Dahr Jamail, *Will to Resist: The Soldiers Who Refuse to Fight in Iraq and Afghanistan.* (Chicago: Haymarket Books, 2011), 47.

courage of the platoon commander to admit to the chain of command their concerns, stopping unethical actions from taking place.

A good litmus test for a leader to judge the professionalism of their unit is how they dress. Military fashion can soon take hold of a unit. In the British Army there is often a pressure to be 'ally' which is slang for wearing the right kind of cool clothing, normally personally purchased in order to be accepted as professional. This can vary from boots to personal equipment, day sacks to clothing and ancillaries. However, research 'studies demonstrate that a nominal clothing manipulation can have effects on the behaviour of the wearer.'[43] The research also found that 'when clothing has symbolic meaning for the wearer, it also affects the wearer's behavior [sic].'[44] In recent conflicts, units have often designed badges that represent them and these have often included a white skull on a black background. Apart from the skull motif being unprofessional, the research noted that 'the color [sic] black is associated with evil and death in many cultures.'[45] The authors concluded that 'the manipulations were designed so that the meaning of the dress cues was salient for the context of the manipulation.'[46] Therefore the leader needs to be aware how their teams dress. As a young officer serving in Northern Ireland in the early 1980s, the battalion commander insisted that only issued uniform would be worn by the soldiers patrolling the streets of Londonderry. Despite the soldiers wanting to be 'ally' it was explained that wearing issued uniform and equipment was an important part of looking smart and professional to the local population that the battalion were there to reassure. In hindsight, it also had an impact on the professional behaviour of the unit. Therefore, the leader needs to consider the manipulation of clothing and ask the question 'is this professional'? If they consider that it is not, they should take action to correct it as this might indicate that the team have taken the first step on the spiral of violence.

Summary

There are four separate types of accountability affecting military personnel, hierarchical, professional, legal and political. These four areas are often in friction, and it can be difficult to balance them. Using the British Army

---

43 Johnson et al, 'Dress, Body and Self: Research in the Social Psychology of Dress', *Fashion and Textiles*, Vol. 1, No. 20 (2014), 8.
44 Ibid., 9.
45 Ibid.
46 Ibid., 10.

Standards of actions being appropriate, lawful and totally professional can assist a leader to keep the accountability types in balance. Remembering the acronym LAP and asking is the action Lawful, Appropriate and Professional can act as a checklist to assist a leader in a stressful decision-making situation. Patrick Bury remembered as a young officer considering decisions in this way and that after Afghanistan, soldiers thanked him 'for taking decisions that ultimately spared them from killing.'[47] Therefore the first key question to ask in a stressful situation and considering accountability is 'Are my actions Lawful, Acceptable and Professional'?

Accountability Case Study – Abu Gharib Abuse

The Abu Gharib abuse was investigated in chapter 4 and the part played by Sergeant Joe Darby to highlight the abuse was explored in chapter 7. It has been stated that the 'Abu Gharib prison scandal has joined accountability and ethics together in a way that other discussions never have.'[48] There have been many examinations into the abuse, but one of the most comprehensive was the official investigation conducted by Major General Antonio Taguba. He produced a report into the actions of the 800th Military Police Brigade at Abu Gharib often known as the *Taguba Report*. The report 'concluded that "sadistic, blatant and wanton criminal abuses" had been inflicted upon detainees.'[49] This case study will analyse the hierarchical, professional, legal and political accountability, or lack of it, and conclude to examine how Sergeant Joe Darby considered accountability.

Hierarchical Accountability

One of the problems earlier highlighted with hierarchical accountability is when units are dispersed and there is little supervision. This was a critical issue in the case of Staff Sergeant Fredrick and the night shift who received little control or supervision from the chain of command. Investigations found that the 'laissez-faire attitude of management coupled with conflicting directives and ambiguous messages about what was acceptable behaviour

---

47 Bury, 'Maintaining Morality at the Tactical Level', 121.
48 Olshfski, 'Accountability and Ethics in the Abu Ghraib Scandal'.
49 'Correct a black mark in US history': former prisoners of Abu Ghraib get day in court', *The Guardian* https://www.theguardian.com/world/ng-interactive/2024/apr/14/abu-ghraib-iraq-torture-abuse.

towards prisoners weakened hierarchal accountability' at Abu Gharib.[50] When Major General Taguba scrutinised the process he found that 'individual Soldiers within the 800th MP Brigade and the 320th Battalion stationed throughout Iraq had very little contact during their tour of duty with either LTC (P) Phillabaum or BG Karpinski.'[51] Lieutenant Colonel Phillabaum was the commander of the 320th Battalion stationed in Abu Ghraib and Brigadier General Karpinski commanded 800th Brigade and was responsible for all prisons in Iraq. The *Taguba Report* went on the state that 'with respect to the 320th MP Battalion, I find that the Battalion Commander, LTC (P) Jerry Phillabaum, was an extremely ineffective commander and leader. ... LTC (P) Phillabaum was also reprimanded for lapses in accountability that resulted in several escapes.'[52] The report went on to explain how Lieutenant Colonel Philabaum was relieved of command due to being found guilty of seven charges one of which was 'failing to properly supervise his soldiers working and "visiting" Tier 1 of the Hard-Site at Abu Ghraib.'[53] Brigadier General Kaplinski was also relieved of command and found guilty of 12 charges including 'failing to ensure that numerous and reported accountability lapses at detention facilities throughout Iraq were corrected.'[54] There is little doubt that the lack of hierarchical accountability by these two officers allowed for an environment in which abuse was conducted without check by the chain of command.

Professional Accountability

The 800th MP Brigade was a reserve formation made up mainly of civilian police and prison staff. Their standard of training and discipline was not the same as full time troops and their immersion into the values of the US Army was not as complete. Research has validated this claim and shown that 'professional accountability was however weakened by the fact that the soldiers were reservists, not professional soldiers.'[55] The *Taguba Report* went further to question not only the unit but the commanders professional action it stated that,

---

50 Olshfski, 'Accountability and Ethics in the Abu Ghraib Scandal'.
51 US Army, Article 15-6 Investigation of the 800th Military Police Brigade, 2004, 43.
52 Ibid., 39.
53 Ibid., 45.
54 Ibid., 44/45.
55 Olshfski, 'Accountability and Ethics in the Abu Ghraib Scandal'.

> On 17 January 2004 BG Karpinski was formally admonished in writing by LTG Sanchez regarding the serious deficiencies in her Brigade. LTG Sanchez found that the performance of the 800th MP Brigade had not met the standards set by the Army ... 'I totally concur with LTG Sanchez' opinion regarding the performance of BG Karpinski and the 800th MP Brigade.[56]

Professional standards had not been set or upheld by the senior leadership and this lack of professional performance and accountability had cascaded throughout the unit. With no trust in the leadership, standards were allowed to slip at the lower levels without check and individuals had little intrinsic motivation to do the best job possible to an acceptable level.

Legal Accountability

There was a major failing in legal accountability in that no training in International Humanitarian Law was directed or conducted either before or during the deployment. Major General Taguba was so shocked by Brigadier General Karpinski's lack of accountability in this area that he references it three times in the report. The *Taguba Report* initially points out that 'there is no evidence that BG Karpinski ever attempted to remind 800th MP Soldiers of the requirements of the Geneva Conventions regarding detainee treatment or took any steps to ensure that such abuse was not repeated.'[57] Later he repeated that 'I could find no evidence that BG Karpinski ever directed corrective training for her soldiers or ensured that MP Soldiers throughout Iraq clearly understood the requirements of the Geneva Conventions relating to the treatment of detainees.'[58] And finally, the report confirmed that one of the 12 charges against Brigadier General Kaplinski was 'failing to ensure that MP Soldiers in the 800th MP Brigade knew, understood, and adhered to the protections afforded to detainees in the Geneva Convention Relative to the Treatment of Prisoners of War.'[59] The *Taguba Report* also points out that there was also no 'evidence that LTC(P) Phillabaum, the commander of the Soldiers involved in the Camp Bucca abuse incident, took any initiative to ensure

---

56 US Army, Article 15-6 Investigation of the 800th Military Police Brigade, 44.
57 Ibid., 20.
58 Ibid., 43.
59 Ibid., 44/45.

his Soldiers were properly trained regarding detainee treatment.'[60] It also highlighted that one of the charges against LTC(P) Phillabaum was 'failing to ensure that Soldiers under his direct command knew and understood the protections afforded to detainees in the Geneva Convention Relative to the Treatment of Prisoners of War.'[61]

There was obviously no consideration of International Humanitarian Law in 800th MP Brigade, which considering its composition of civilian police and prison staff is surprising. However, it has been pointed out that this came from the top and that 'the military leadership failed with respect to legal accountability.'[62] Despite this lack of example and direction when looking at the outcome of the situation it has been noted that 'only a handful of lower-rank soldiers faced military trials; no military or political leaders, or private contractors, were held legally accountable for what happened at Abu Ghraib.'[63] It could be argued that the soldiers knew what they were doing was illegal, but in the ambiguous environment of Abu Gharib and without defined rules they conducted abuse unchecked.

Political Accountability

The abuse at Abu Gharib conducted by a handful of junior soldiers on the night shift in Iraq quickly became an international story. A commentator on these events noted that 'we live in the era of the strategic corporal. Immoral behavior [sic] by even the lowest ranking soldier can have a strategic effect, as witnessed by the impact of the images of Private Lynndie England, a "strategic private," at Abu Ghraib prison in Iraq.'[64] There was however not much focus or detailed research into the political ramifications of the actions of 800th MP Brigade, with most research concentrating on the US Government of President George W. Bush. The conclusion of many researchers was that 'although many of those in the [Bush] administration knew about detainee mistreatment they chose not to do anything about it.'[65] There was a belief that many political leaders believed that admitting to an awareness of illegal

---

60 Ibid., 20.
61 Ibid., 45.
62 Olshfski, 'Accountability and Ethics in the Abu Ghraib Scandal'.
63 The Guardian, 'Correct a black mark in US history': former prisoners of Abu Ghraib get day in court'.
64 Paul Robinson, 'Ethics Training and Development in the Military', *Parameters*, Vol. 37, No. 1 (2007), 25.
65 Olshfski, 'Accountability and Ethics in the Abu Ghraib Scandal'.

activities and taking no action would be dangerous as this avoidance of accountability could leave them open to international prosecution. However, once the story of what had happened was made known by Sergeant Joe Darby, the political fallout for the Bush administration was huge.

Actions of Sergeant Joe Darby

The actions of Sergeant Joe Darby have been explained in chapter 7. He not only stopped the abuse but enabled those responsible at all levels being held to account. It was however not an easy decision, when he received an award for having the courage to pass on the information which exposed the abuse, he stated that 'I was faced with the toughest decision. On one hand, I had my morals and the morals of my country. On the other, I had my comrades, my brothers in arms' however he concluded that he was 'proud of my decisions to put the values of my country and its reputation ahead of everything else.'[66] Presenting the award, Senator Edward Kennedy expressed how 'Joseph Darby courageously refused to remain silent, and in so doing he embodied the best of our American values when he said "enough" at Abu Ghraib. He is a true Profile in Courage.'[67] In a subsequent interview Sergeant Darby recalled how the abuse 'violated everything that I personally believed in and everything that I had been taught about the rules of war.'[68] Sergeant Darby asked himself are these actions lawful, acceptable and professional? Darby's ultimate behaviour reflected an individual's ability to act when they believe that legal and professional accountability have been broken and hierarchical accountability has not been effective in controlling it.

---

66 John F. Kennedy Courage Awards 2005, , May 16, 2005, https://www.jfklibrary.org/events-and-awards/profile-in-courage-award/award-recipients/joseph-darby-2005.
67 Ibid.
68 Ibid.

10

S-CALM:
L – Leadership

Introduction

In chapter one I mentioned a soldier that was killed in my first command in Northern Ireland. 30 years after the soldier's death, the platoon decided to meet at the grave in Cambridge and I was asked to attend. As I arrived, I saw two familiar faces in the car park, they were soldiers from the platoon I had not seen for 30 years. We shook hands, and they said, "Hello Sir". I said to them, "Please call me Dennis", but they replied, "We didn't know that was your name". For the two years that we had served in our tiny patrol base in Northern Ireland, I had been "Sir" in the base or "Boss" when we had been on patrol. We had lived cheek by jowl in two portacabins on operations for six months over a two-year period and I knew every detail about each one of them. And yet despite this closeness I was their leader and they were my team. I was 'in the platoon' but not 'of the platoon' and knew that I would never be as close to them as they were to each other. Therefore, during those two years I had managed to retain a psychological distance that allowed me to intervene and be the conscience of the platoon when difficult decisions were required, this is what I refer to as the loneliness of command. I alone in the platoon would take the difficult ethical decisions and stop unethical actions.

The well-known leadership academic Ralph Stogdill wrote that 'there are almost as many different definitions of leadership as there are persons who have attempted to define the concept.'[1] However when it comes to dealing with ethical leadership it is not the theory or style which is important,

---

1 Ralph Stogdill, *Handbook of Leadership*. (London: Collier Macmillan, 1974), 7.

what is of paramount importance is that those in command display courageous moral leadership in difficult situations. To do this, leaders must be consistent in their supervision, attitudes and activities, in order to build trust and mutual respect with their teams. They will need to both inspire and direct their teams in difficult situations, and this requires the application of both transformational and transactional leadership techniques. British Army doctrine states how leaders 'must be authentic and genuine, and lead by example. Leaders are role models; their behaviours and values are judged and assimilated by those they lead.'[2] Nevertheless there are times when a leader will feel under considerable pressure to conform to the group. It is in these circumstances that the loneliness of command comes into its own. The idea captures how there are times when a leader is required to rely on their own council and act for the good of the team.

The beginning of this chapter is broken down into three sections which deal with the three aspects of command, leadership and management in difficult situations. In each section two key aspects are investigated. Under command the concept of the loneliness of command and the difference between commission and omission will be examined. The section on leadership covers the requirement to be the exemplar who sets the moral tone of an organisation and the necessity to gain and maintain followers' trust. In the management section the requirement to set out followers' expectations and the need to apply transactional leadership in the shape of rewards and punishments to motivate ethical behaviour is discussed. There is a final section which explores the idea of reasonable challenge and investigates how to both make and receive challenges to authority. Each of these sections concludes with a real-life example from Patrick Bury, who was a second lieutenant in combat operations in Afghanistan. Following a summary, there is a leadership case study which explores the ethical leadership displayed by the General Sir Patrick Sanders over the actions of the 3rd Battalion the Parachute Regiment.

Command

The first of the three areas to be explored is command. Command comes from the old French part of the English language 'comander' which basically means to order. Command in the British military system is about holding an appointment, British Army doctrine defines it as follows:

---

2 MOD, *Army Leadership Doctrine*, 2-9.

> Command is the authority vested in an individual for the direction, coordination and control of military forces. It has a legal and constitutional status, codified for the Army in the Queen's Regulations. It is delegated to a commander by a higher authority that gives direction and assigns forces to accomplish a mission. The exercise of command is the process by which a commander makes decisions, impresses their will on, and transmits intentions to subordinates. It entails authority, responsibility, and accountability.[3]

The final sentence of the definition is an important aspect of command in that the appointed commander not only has authority, but the corresponding responsibility and accountability that goes with it. Exploring British Army ethics, Philip McComack summarised this point when he commented 'the creation and maintenance of an appropriate ethical climate within a unit is a function of command'.[4] The two aspects of loneliness in command and commission and omission will now be considered.

Loneliness of Command

When speaking to trainee officers at the Royal Military Academy Sandhurst, Field Marshal The Viscount Slim explained how 'there will come a pause when everyone will look to you. They will look to you for leadership. Believe me, when that happens, you will feel very lonely.'[5] At the beginning of this chapter, I explained how I maintained a relationship where I was 'in' the team but not 'of' the team and managed to keep a separation which allowed for successful ethical decision making. This concept of psychological distancing, of being alone in the group is termed the loneliness of command. Psychological distance has been defined as the 'psychological effects of actual and perceived … differences between the supervisor and subordinate.'[6] The ability to conduct this type of relationship is built on mutual respect and an understanding of

---

3   MOD, *Army Leadership Doctrine*, 1-4.
4   Phillip McCormack, *Preparing Professional Military Forces to Face Ethical Challenges in Future Military Operations*, EuroISME Presentation, 2015, 8.
5   Slim, W, 'Address by Field Marshal The Viscount Slim On 14 October 1952 to Officer Cadets of The Royal Military Academy Sandhurst', https://www.pnbhs.school.nz/wp-content/uploads/2015/11/Slim.pdf.
6   B. Napier and G. Ferris, 'Distance in organizations', *Human Resource Management Review*, Vol. 3, No. 4 (1993), 328.

the different roles and responsibilities within a team. To judge officer trainees understanding of the need to have a psychological air gap with their team, the RMAS 2023 research survey asked the question: It is OK for a platoon commander to use the first names of their soldiers? For the cohort with no previous military training there was a statistically significant shift between their first response on arrival with 24 per cent agreeing to the final response which had 49 per cent in agreement. The research noted that 'this result suggests that Ocdts [Officer Cadets] arrived with a view that there needs to be an air gap between officers and soldiers, but that this changed whilst at Sandhurst to nearly half considering that this behaviour was acceptable.'[7] The responses from all trainees at the end of training was similar for British trainees with 53 per cent agreeing that it was acceptable to use soldiers first names and 31 per cent stating that this was not acceptable. For the 'international Ocdts 47% disagreed, suggesting a traditional belief in leader/follower relationships.'[8] This demonstrates a lack of understanding in training for the need to retain a healthy command relationship so that when required they will be able to make the key loneliness of command decisions. Nevertheless, they were encouraged to develop a leadership style built on mutual respect which needs to be practiced as part of the normal command relationship. It cannot be turned on when required and then turned off again. In this vein, the Australian Defence Force explain how the 'true measure of a leader is not where you stand in moments of comfort, but where you stand in moments of challenge and difficulty. Sometimes that will mean standing alone.'[9] It was during one of these periods of difficulty in the Falkland's War that Captain David Hart Dyke wrote home that 'the loneliness of command, especially in difficult times, is quite a strain – though I know I shall cope all right.'[10]

Commission and Omission

The second aspect is understanding that unethical acts are conducted by either commission or omission by a commander. Acts of commission are well recognised. In these situations perpetrators take an active decision to act in an unethical way and often use their authority to conduct these actions. An example of an act of commission was the actions of Lieutenant Calley during

---

7 Vincent and Muhl-Richardson, *An Analysis of the Effectiveness of the S-CALM Model of Ethical Leadership*, 18.
8 Ibid.
9 ADF, *Leadership*, 13.
10 Hart Dyke, *Four Weeks in May*, 31.

the My Lai massacre. However, it has been noted that 'there are more ways to bring about harm through omission than through commission.'[11] This is when acts of omission, which are the opposite to commission come in. Acts of omission are ones in which unethical acts are committed by inaction and passive behaviour such as when people succumb to the bystander effect. An example of an act of omission might be the lack of supervision demonstrated by the leadership at Abu Gharib. In many cases 'omission comes in the form of choosing status quo or perhaps just deferring a decision or not even considering possible options. In these cases, the lack of a decision is the same thing as a decision, just a bad one.'[12] The need for a leader to make a decision when faced with an ethical dilemma is the critical point, as to not do so can allow for the commission of unethical actions. In many of the case studies in this book leaders did not intend for unethical actions to take place and research has shown that 'harmful omissions are typically less intentional than commissions.'[13] A leader must be prepared to take proactive, positive action. Patrick Burry recalled how 'at some stage in your early career, you will be faced with a situation where you will know what is right to do and somebody under your command will be challenging this knowledge. Do you shirk your responsibility for the sake of popularity.'[14] He goes on to explain that this is the moment when an ethical leader must stand up, enact the loneliness of command, and make a decision that will ensure that the team conducts ethical actions.

Leadership

The second area is leadership. The word leader comes from the Anglo-Saxon ledian/lædan, which means a path or journey. Although it has already been pointed out that there are too many definitions of leadership, it is important before continuing to identify what militaries mean when they talk of good leadership. British Army doctrine gives a definition of leadership as being 'a human endeavour. It is persuasion, compulsion and example that inspire others into action. Given that leadership does not require designated authority, anyone can lead. All commanders, those granted with authority, however, are

---

[11] Mark Spranca, Elisa Minska and Jonathan Baron, 'Omission and Commission in Judgment and Choice', *Journal of Experimental Social Psychology*, Vol. 27 (1991), 103.
[12] Rob Thomas, 'Acts of Omission vs. Commission', 2018, https://robdthomas.medium.com/acts-of-omission-vs-commission-4d494a6b0ec8.
[13] Spranca, Minska and Baron, 'Omission and Commission in Judgment and Choice', 101.
[14] Bury, 'Maintaining Morality at the Tactical Level', 121.

expected to lead.'[15] It continues 'good leadership instils purpose, confidence, motivation, and ethical conduct.'[16] The US Army Special Forces define being a moral person as referring to 'your own personal ethical behavior [sic] and is, of course, absolutely critical to developing a reputation as an ethical leader'.[17] There are two areas of leadership that will be investigated, setting the moral tone and the requirement for mutual trust.

Set The Moral Tone, Be The Exemplar

The first key element in displaying ethical leadership is the requirement to be an example to the team and thus set the moral tone of the unit. Being the exemplar is a full-time task and followers will have high expectations of their leaders and use them as a role model for their own actions. Research had demonstrated that the maintenance of 'high ethical standards applied to both ethical role models' public lives in the workplace and their personal lives outside the workplace.'[18] Therefore the task does not stop when a leader removes their uniform. It is however important as Canadian research established that armed forces members 'found that ethical leaders encourage ethical behaviour among their subordinates' it also noted that 'the ethical climate created within a unit by those in command before an operational deployment is likely to be replicated on operations.'[19] The Canadian research concluded that,

> Leaders thus have an important role in shaping an organization's ethical culture. In addition to modelling ethical behaviour, other aspects of ethical culture that influence ethical behaviour are commitment from leaders and employees to act ethically, to be transparent, to be open to discussing ethically challenging situations, and to discipline personnel who violate ethical standards.[20]

---

15 MOD, *Army Leadership Doctrine*, 1-4.
16 MOD, *ADP Land Operations, Part 2: The Application of Land Power*. (Warminster: Land Warfare Centre, 2022), 1-5.
17 US Army, *A Special Operations Forces Ethics Field Guide*, 20.
18 Gary Weaver, Linda Trevino and Bradley Agle, 'Ethical Role Models in Organizations', *Organizational Dynamics*, Vol. 34 No. 4 (2005), 319.
19 Deanna Messervey, Waylon Dean, Elizabeth Nelson and Jennifer Peach, 'Making Moral Decisions Under Stress: A Revised Model for Defence', *Canadian Military Journal*, Vol. 21, No. 2 (2021), 43.
20 Ibid.

Military doctrines expand on the idea of good role models and being the exemplar as an important theme of leadership. British Army doctrine notes 'leaders have an important role in promoting self-discipline through their own personal example.'[21] It then adds that 'you cannot lead people beyond where you can and are willing to go yourself. All leaders are role models, and as an Army leader you must actively demonstrate our Values in everything you do.'[22] The US Army has a similar line stating that 'leaders are ethical standard-bearers for the organization, responsible for establishing and maintaining a professional climate wherein all are expected to live by and uphold the Army Values.'[23] It concludes that 'it is critical that leaders rely on positive behaviors [sic] to influence others and achieve results.'[24] The Australian Defence Force concur stating that 'Good role models demonstrate the behaviours expected from those within the team.'[25] They also concluded that 'all leaders are responsible not only for their own actions, but also for setting the ethical environment in their team.'[26] Therefore it is clear that a military leader is expected to be the role model for their unit and that the actions they conduct, both on and off duty, will establish the moral tone of their unit. Patrick Bury recalled that 'setting the example is also extremely important in building trust and respect ... I have seen competent commanders instantly lose the respect, and therefore trust, of their soldiers by ignoring the same safety induced hardships that their soldiers are subjected to.'[27]

Gain and Maintain Trust

Trust is a leader's central commodity and without it leadership is extremely difficult. It is vital followers trust their leaders' abilities and have confidence in their capability to make ethical decisions. Sandra Reinke believed that trust is developed and displayed by an ethical leader in four dimensions, these are 'fostering open communications, listening, being competent and predictable, [and] caring.'[28] Most of these dimensions of trust can be achieved by a leader demonstrating emotional intelligence, having integrity

---

21 MOD, *Army Leadership Doctrine*, 2-5.
22 MOD, *The Army Leadership Code*, 16.
23 US Army, *Army Leadership*, 6-5.
24 US Army, *ADP 6-22*, 8-8.
25 ADF, *Military Ethics*, 38.
26 Ibid., 5.
27 Bury, "Maintaining Morality at the Tactical Level", 121.
28 Saundra Reinke, 'Service Before Self: Towards A Theory Of Servant-Leadership', *Global Virtue Ethics Review*, Vol. 5, No. 3 (2004), 41.

and empathy with their teams and therefore seeing themselves as others see them, the so called 'Looking Glass Leadership'. Once a leader can observe their decisions and actions from the viewpoint of their team, it enables them to ensure that they are communicating their intent clearly. The Australian Defence Force doctrine states that 'teams that trust and respect each other, work collaboratively and encourage open communication usually develop sound ethical cultures.'[29] The need for competence and predictability is also important. Followers want to know where they stand with a leader and that their reactions will be consistent. Research has shown that leaders who 'exhibit predictability and consistency within an open and ethical climate ... build the "culture of trust"' in their team.[30] It takes time for a leader to build trust and gain the confidence of their team that their ethical decisions are nearly always right. However, a leader can lose trust and confidence very quickly by violating their followers' expectations in any one of the four dimensions, but especially predictability. British Army doctrine highlights how 'all ranks must avoid behaviour that risks degrading their professional ability or which may undermine morale by damaging the trust and respect that exists between teams and individuals who depend on each other.'[31]

Finally, there is an unwritten psychological contract between leader and followers, 'a perceived implicit and reciprocal exchange between the employer and employee.'[32] This unwritten agreement is that the leader will care for their team and in return the team will give their support. British Army doctrine explains how 'trust and mutual understanding between leaders and subordinates must be implicit.'[33] It goes on to explain how 'trust in the Chain of Command is also key, and demands integrity from those in positions of authority.'[34] The need for leader integrity is echoed by the US Army doctrine which asks 'what makes an effective leader: a person of integrity who builds trust' and also by the Australian Defence Force which concludes that – 'integrity is necessary to establish trust. It is not the only ingredient, but its absence makes trust impossible.'[35] Therefore it is an imperative for a leader to first build and then maintain the trust of their team.

---

29 ADF, *Military Ethics*, 38.
30 Reinke, 'Service Before Self', 37.
31 MOD, *The Army Leadership Code*, 10.
32 David Hall, Stephen Pilbeam and Marjorie Corbridge, *Contemporary Themes in Strategic People Management: A case-based Approach*. (Basingstoke: Palgrave Macmillan, 2013), 56.
33 MOD, *Army Leadership Doctrine*, 1-5.
34 MOD, *The Army Leadership Code*, 9.
35 US Army, *ADP 6-22*, v. ADF, *Leadership*, 13.

Patrick Bury commented that 'your soldiers must trust you to do the right thing. They must trust your decisions. They must respect you.'[36]

Management

The last of the three areas is management. This comes from the Latin manus, which means to handle. Famously, Field Marshal Sir William Slim suggested the relationship between leadership and management; he said that 'leadership is of the spirit compounded of personality and vision; its practice is an art. Management is of the mind, more a matter of accurate calculation, of statistics, of methods, timetables, and routine; its practice is a science.'[37] This basic concept of the difference between the two ideas being one an art and another a science has passed into the normative understanding of the different aspects of command. British Army leadership doctrine defines management as follows,

> Management relates to the systems, processes and mechanisms for the control and allocation of resources (staff, equipment, financial resources, etc). It uses organisational systems and processes to minimise risk and to achieve results in the most efficient manner possible. In essence, sound management contributes to the building of trust and loyalty that facilitates effective leadership.[38]

The two key areas of management that are essential to ethical leadership are the requirement to set expectations for the team and need to be able to apply rewards and punishments to motivate ethical behaviour.

Setting Out Expectations

There is a requirement for leaders to ensure that their teams understand what is expected of them. The Australian Defence Force believe that 'setting expectations, monitoring performance and addressing any breaches provides subordinates a clear understanding of what is expected of them.'[39] Following the Falkland's War Captain Hart Dyke recalled that 'it was simply my job

---

36 Bury, 'Maintaining Morality', 121.
37 William Slim, Address to the Adelaide Division of the Australian Institute of Management, 4 April 1957, *Australian Army Journal 1957* (1957), 7.
38 MOD, *Army Leadership Doctrine*, 1-5.
39 ADF, *Military Ethics*, 38.

to provide strong leadership and to ensure my crew followed me willingly, whatever the hardships, and I found this could be more easily achieved by showing care and respect and keeping them as well informed as I could about the situation.'[40] Ensuring that the team understands what is expected of them is done in the long term by establishing and communicating a clear vision and in the short term by giving an unambiguous intent on what is required. British Army doctrine says of vision that leaders 'must provide clear and unifying purpose, generating a sense of team cohesion and direction. The vision can be expressed by both communication and by role modelling.'[41] It also emphasises how it is a 'leader's responsibility to provide a vision of shared goals so that individuals and the team are inspired.'[42] An unambiguous intent in a stressful situation is also vital. When Lieutenant Kallop arrived at the scene of the Haditha massacre and gave the ambiguous order "to take the house", it ended in the killing of 19 civilians. Therefore, despite the stress of the situation, a leader must ensure that their orders are clear and ethical in nature. The US Army summarise this thinking by stating that 'the ultimate responsibility to create and maintain an ethical climate rests with the leader.'[43] During Patrick Bury's deployment in Afghanistan he was involved in many combat situations and had to give clear direction, he recalled that 'my soldiers thanked me after the tour for taking decisions that ultimately spared them from killing, even when it would have been allowed in our ROE [Rules of Engagement].'[44]

Apply Transactional Rewards and Punishments to Motivate Ethical Behaviour

Ethical leadership is the only transformational leadership style that has a transactional element to it. This is due to the requirement for ethical leaders to hold 'their subordinates accountable to high ethical standards.'[45] To meet these ethical standards 'leaders use reward and punishment systems to influence followers' behaviour.'[46] In doing this it is essential that 'reward

---

40 Hart Dyke, *Four Weeks in May*, 222.
41 MOD, *The Army Leadership Code*, 14.
42 MOD, *Army Leadership Doctrine*, 5-2.
43 US Army, *Army Leadership*, 6-5.
44 Bury, 'Maintaining Morality', 121.
45 Weaver, Trevino and Agle, 'Ethical Roles Models', 322.
46 David M. Mayer, Karl Aquino, Rebecca L. Greenbaum And Maribeth Kuenzi, 'Who Displays Ethical Leadership, And Why Does It Matter? An Examination of Antecedents And Consequences Of Ethical Leadership', *The Academy of Management Journal*, Vol. 55, No. 1 (2012), 167.

and punishment should be linked to desired standards of behaviour and performance.'[47] Many militaries are efficient at punishing people for breaking their ethical code, but slow to reward those that uphold it. However, in order to change behaviour and motivate people there is a requirement for both. Motivation can be divided into extrinsic and intrinsic forms. Extrinsic motivation is the application of external factors like rewards and punishments and tend to be focused on either an outcome or avoidance of punishment. Intrinsic motivation is from within and is focused on achieving a positive outcome in activities that are their own reward. John Adair in *Effective Motivation* studied this mix of extrinsic and intrinsic motivation and stated that for most people '50% of motivation comes from within a person, and 50% from his or her environment, especially from the leadership encountered therein.'[48] Many leaders are exceptionally intrinsically motivated and it can be difficult for them to consider that they need to employ a high level of extrinsic motivation for their followers but this is important if they are to change their behaviours. The rewards do not have to be large or financial to have an effect and sometimes just telling someone they are doing a good job can work as 'it is human nature to enjoy being praised, and reward recognises effort, inspiring further endeavour and motivation to do even better.'[49] However this book has outlined some case studies of where rewards for ethical actions have been awarded, such as the Military Cross to Major Richard Westley for leadership in Bosnia, the award of the soldiers Medal to Warrant Officer Thompson for stepping in and stopping the My Lai massacre and Sergeant Joe Darby receiving the John F. Kennedy Profile in Courage Award for exposing the abuses at Abu Ghraib. Patrick Bury identified that 'you must build and maintain close working relationships with all of your soldiers. You must know your men, their personalities and their problems and above all how to motivate them.'[50]

It might be questioned why punishment is a motivator for ethical decision-making. However, the application of fair discipline which is suitable for the crime committed is an important aspect of morale for the military. The overarching British Army doctrine for operations states that 'discipline is essential and is part of the ethical foundation and unique culture of the

---

47 MOD, *Army Leadership Doctrine*, 5-4.
48 John Adair, Action Centred Leadership, 2015. https://www.valuing-your-talent-framework.com/sites/default/files/resources/THK-032%20John%20Adair.pdf.
49 MOD, *The Army Leadership Code*, 17.
50 Bury, 'Maintaining Morality', 121.

force.'⁵¹ Nevertheless in some of the case studies examined there has been a lack of punishment of the perpetrators. In the case of Batang Kali, none of the Scots Guards who killed 24 civilians were ever brought to justice and in most of the cases of incidents conducted by the US forces in Iraq, except for the Abu Gharib abuses, US personnel had no disciplinary sanctions held against them. Even in some high-profile cases, such as Lieutenant Calley at My Lai and Sergeant Blackman in Afghanistan, although there was an initial attempt at discipline, neither fulfilled their initial punishments due to a lack of government will. These lapses in holding perpetrators to account can have a toxic effect on organisations. Research had demonstrated that 'if unethical decision making is rewarded, then higher incidence of unethical behavior [sic] is likely to occur.' The research concluded that 'in situations in which potential reinforcement for unethical behavior [sic] does exist, it should be accompanied by strong punishment (or threat of punishment).'⁵² It is for this reason that most militaries have identified the need to have an espoused policy of dealing robustly with those that break their ethical code. The British Army doctrine states that 'the application of Discipline, regardless of rank is crucial to correct failings and punish transgressions.'⁵³ Whilst the US Army insist that 'an effective leader instils discipline by training to standard, using rewards and corrective actions judiciously, instilling confidence, building trust among team members.'⁵⁴ Finally the Australian Defence Force explain their use of reward and punishment and how 'from time to time you will have to deal with unacceptable behaviour. We have a discipline system whose purpose includes protecting individuals and teams from such behaviour… There is a formal honours and awards system. Use it to reward outstanding achievement.'⁵⁵ Judging when to apply both reward and punishment is an important skill for a leader as the unjust application of either will affect morale. It has been noted however that 'leaders who are high in moral identity internalization are more likely to pay attention to, correct, and punish unethical behaviors [sic].'⁵⁶ Patrick Bury recalled how not all punishments had to be administered by the leader and that 'for minor breaches, the group dynamic is a useful way of chastising those who act

---

51 MOD, *ADP Land Operations*, Part 2, 1-5.
52 Harvey Hegarty and Henry Sims, 'Some determinants of unethical decision behavior – An experiment', *Journal of Applied Psychology*, Vol. 63, No. 4 (1978), 456.
53 MOD, *The Army Leadership Code*, 17.
54 US Army, *Army Leadership*, 5-7.
55 ADF, *Leadership*, 19-20.
56 Mayer et al, 'Who Displays Ethical Leadership', 153.

inappropriately. For major breaches use your chain of command. Do not let indiscipline or laziness corrode your unit.'[57]

Challenge

An essential part of being an ethical leader is having the ability to accept a challenge from the team without seeing it as a threat to their authority. Allowing followers this right to question is known as intelligence disobedience, a term first coined in 1936 to describe how guide dogs were trained to override owner's commands at times of danger. For intelligent disobedience to work correctly, it is not just the leader that has to be able to accept a challenge, it also calls for a mature attitude amongst the team. This has been described as the requirement for followers to examine 'ways in which a trusted relationship can be developed with leaders that will support speaking candidly and acting with integrity.'[58] The ability to speak candidly can only be achieved when followers believe that there is a high degree of psychological safety in the team. Psychological safety has been defined as a feeling of being 'able to show and employ one's self without fear of negative consequences to self-image, status, or career.'[59] A psychologically safe team can be created 'if the leader is supportive, coaching-oriented, and has non-defensive responses to questions and challenges' in this situation team 'members are likely to conclude that the team constitutes a safe environment.'[60] In teams with a safe environment 'members feel safe to speak up about issues with colleagues to improve team performance, and there is evidence that it is related to individual wellbeing, team functioning and enhanced performance, including safer and ethical practices as well as learning from mistakes.'[61] British Army doctrine on psychological safety states that leaders must 'foster an environment which enables challenge to occur … a true challenge culture, enabled by psychological safety, gives an organisation its competitive advantage.'[62]

---

57 Bury, 'Maintaining Morality', 122.
58 Ira Chaleff, *Intelligent Disobedience: Doing Right When What You're Told To Do Is Wrong*. (Oakland: Berrett-Koehler Publishers, 2015), 185.
59 William Kahn, 'Psychological Conditions of Personal Engagement and Disengagement at Work', *Academy of Management Journal*, Vol. 33, No. 4 (1990), 708.
60 Amy Edmondson, 'Psychological Safety and Learning Behavior in Work Teams', *Administrative Science Quarterly*, Vol. 44, No. 2 (1999), 356.
61 NATO, *Leader Development for NATO Multinational Military Operations*, STO-TR-HFM-286. (Boston Spa: NATO, 2022), 115.
62 MOD, *British Army Followership Doctrine*. (Camberley: The Centre for Army Leadership, 2023), 16.

Challenge in the Military

Should a leader in the military accept a challenge to their authority? The British Army believe that this is a vital skill and that all 'leaders must create a culture in which followers recognise that constructive challenge is an essential behaviour within a team and is a form of loyalty in itself.'[63] It does continue that for challenge to work good judgement is required by both the leader and follower and 'for the relationship to function optimally, a leader must create an atmosphere of trust, empowerment, and inclusivity.'[64] The British Army has therefore developed what is called a challenge spectrum. At the left end is 'Valued Feedback' which has low impact and requires little moral courage. Valued feedback concerns 'the provision of information or a perspective on an action, event or performance' and is normal in high performing teams. In the centre of the spectrum is 'Constructive or Reasonable Challenge' which is the ability to 'question to hold to account, or to present an alternative to the norm.' Finally at the right of the spectrum there is 'Intelligent Disobedience', which 'entails the potential need and reasonable freedoms to contravene a direct order or instruction.'[65] The US Army do not have a defined structure, but in chapter 7 it was explained how they have the right to refuse an unethical order. The Australian Defence Force also have no formal structure but assert that individuals be 'willing to speak our minds because we are confident that our views matter and that sometimes leaders and common wisdom need to be challenged.'[66] To complement the British Army challenge spectrum, the UK MOD offer advice on how to give and receive a challenge. The MOD's *The Good Operation* describes an environment in which challenge is expected and accepted as being important. It outlines how 'people should be receptive to reasonable challenge and assume that it is provided with the best of intentions.' It continues that 'those offering challenge should know how to do so effectively. Challenge isn't about proving someone right or wrong; rather it's about highlighting and exploring alternative options.'[67] The concept of a devil's advocate or what the British Army calls a 'red team' has already been covered and it is vital that anyone who is appointed to this role fully embraces the best practice for offering a challenge. The final page

---

63 MOD, *Army Leadership*, 2-6.
64 MOD, *Army Leadership Doctrine*, 3-5.
65 MOD, *A British Army Followership Doctrine Note*, 15/16.
66 ADF, *Leadership*, 2.
67 MOD, *The Good Operation: A handbook for those involved in operational policy and its implementation*. (London: HMSO, 2017), 62.

For those offering challenge	For those receiving challenge
Make challenge with courtesy and politeness.	Not take it personally - the challenge isn't about you, it's about the issue at hand.
Be prepared to explain the logic and reasoning behind your alternative view and provide evidence. Keep your challenge concise and relevant to the issue at hand.	Make it known that you welcome reasonable challenge, and create space in the way you run your business to receive it. Recognise that challenge might result in change.
Think about the interpersonal dynamics. Keep it professional - it's the issue you're challenging, not the person. Be respectful to the approach from which you are differing.	Seek real diversity of thought, not just shades of mainstream thinking.
	Give staff the opportunity to fully articulate different views and give them credit for doing so. And remember that the person challenging shouldn't be expected to have a solution there and then.
Choose your moment and your medium. A one-to-one discussion or a smaller team meeting may be more appropriate than a big meeting at which positions are being taken and decisions are expected; a gently probing conversation or email is better than a confrontational one.	
	Demonstrate that you are giving serious thought to the challenge being offered - do not dismiss it out of hand and make sure people aren't just telling you what you want to hear.
Raise issues in a timely manner. Don't leave your challenge too late in the process, when changing course could be too difficult.	Respond respectfully - never belittle someone's view, and never (even after the event) sideline those offering it.
Accept if the eventual decision remains unchanged – a decision has to be taken once all reasonable challenge has been considered. Only in cases where regularity or propriety have not been observed should you need to turn to the Department's whistleblowing process.	If you do not accept the challenge, explain your reasoning, including supporting evidence when necessary.
	Encourage the use of evidence from beyond the immediate organisation, think tanks, academia and other sources.
	Support both junior colleagues and peers to raise a challenge with more senior colleagues'.

Table 10.1 Offering and Receiving a Challenge

of *The Good Operation* explains how to offer a challenge and if challenged how to accept it. This guidance has been summarised in Table 10.1.

Offering advice to future young officers, Patrick Bury recommended that they 'watch for signs of mental degradation or unacceptable callousness creeping in. Challenge any unacceptable behaviour the moment it occurs and do not let it occur again.'[68]

---

[68] Bury, 'Maintaining morality at the tactical Level', 122.

Summary

Ethical leadership needs to be reflected in command, leadership and management. Many of the notions mentioned, such as gaining and maintaining trust, setting expectations and applying rewards and punishments are important to create the collect environment where psychological safety is the norm and a challenge culture is created. However, one of the essential concepts is the requirement to be aware of omission as a leader and consider the consequences of decisions and inactions. This leads to the crucial requirement to make the important decisions and the second key question that a leader must ask which is 'do I need to apply the loneliness of Command? Patrick Bury recalled that 'from my own experiences as an infantry platoon commander, and of combat operations in Afghanistan, morality was something I didn't really understand until I was faced with killing and death on a daily basis. Then, quickly and unexpectedly, it became a critical factor ... in my decision-making process.'[69]

Leadership Case Study – General Sir Patrick Sanders and the Actions of the 3rd Battalion, The Parachute Regiment

In June 2022 a video went viral on social media which appeared to show eight members of the 3rd Battalion, The Parachute Regiment (3 PARA) participating in an orgy with a civilian woman at Merville Barracks in Colchester while dozens of others watched. General Sir Patrick Sanders had just been appointed as the Chief of the General Staff and was outraged at the video. He ordered an investigation by the Royal Military Police, but this found that the sex had been consensual and therefore established that no crime had been committed. Nevertheless there was public outrage at the immoral behaviour causing the Minister for the Armed Forces, Mr James Heappey, to say on television that 'I am aware of that clip and it is disgraceful and people sort of say what, it was consensual, that doesn't matter.'[70] It seemed that nothing could be done despite General Sanders stating that he had 'temporarily lost trust in the Parachute Regiment's 3rd Battalion.'[71] He knew that this was a time to exercise the loneliness of command and

---

69 Ibid., 120.
70 'Colchester Army barracks 'sex videos' are disgraceful says minister', BBC News, 9 June 2022, https://www.bbc.co.uk/news/uk-england-essex-61743173.
71 'Parachute Regiment Balkans deployment cancelled after sex videos', BBC News, 18 June 2022, https://www.bbc.co.uk/news/uk-england-essex-61851456.

decided on an action that would punish 3 PARA without resorting to formal discipline. He wrote an open letter to the battalion which was intended to be transmitted to the whole Army to send an important message to them all about the ethical standards he expected. It began 'my message to the Army is clear: Our licence to operate is founded on trust and confidence and we must hold ourselves to the highest standards.'[72] Although there was no disciplinary action that could be taken, he gave orders that 3 PARA were to be removed from their planned deployment to the Balkans. In September 2022, 400 soldiers from 3 PARA were due to deploy to Bosnia and Kosovo for over a month, finally participating in a ten-day NATO exercise. The deployment would have gained them an operational medal and additional deployment pay. This unusual punishment was taken badly by the soldiers of 3 PARA. The colonel commandant of the Parachute Regiment, Lieutenant General Andrew Harrison, wrote a letter to the battalion noting that they would 'be devastated by the lost opportunity to serve on operations once again in the Balkans'. However, he hoped that they would 'reflect on and where required adjust behaviours and culture that undermines the reputation and operational effectiveness of an exceptional battalion.'[73] General Sanders's bold action demonstrated that he was a leader of high moral standing who was not afraid to make unpopular decisions in order to ensure that the ethical standards of the British Army were upheld.

---

[72] 'Army cancels Balkans deployment after paratroopers orgy video', The Independent, 18 June 2022, https://www.independent.co.uk/news/uk/army-orgy-video-kosovo-bosnia-b2104022.html.
[73] BBC News, 'Parachute Regiment Balkans deployment cancelled after sex videos'.

11

S-CALM:
M – Moral Compass

Introduction

As Commanding Officer of an infantry battalion, I was given some direction by the Brigade Commander. The direction was to take some action with my battalion that I did not consider to be morally correct. I pondered on this direction, which was not a direct order but clearly an activity that he wanted to happen. I believed that the action was not ethical, and I said that I would not be prepared to conduct it with my battalion. He then applied some pressure on me by pointing out that other Commanding Officers in the brigade had agreed to do it and that I was out of step with my peers. I reconsidered the direction, but also remembered the adage that goes 'Wrong is wrong even if everyone is doing it. Right is right even if no one is doing it'. Therefore, I said that regrettably I still would not comply with the request. This didn't go down well, as at that time the idea of intelligent disobedience was not common in the British Army and it was seen as insubordination. Although no direct action was taken against me at this time, it is likely that this conversation ended my chance of promotion from the role, however my conscience was clear and the battalion was not forced to conduct an unethical action.

There are three types of ethical theory in the western tradition, which are known as normative ethics. They provide the different practical means of determining a moral course of action. They are consequentialism or utilitarianism, deontology and virtue ethics. The first theory is consequentialism or utilitarianism which is focussed on the consequences of a particular action and dictates that a morally right action produces a good outcome. The mantra for this theory is that the ends justify the means.

Utilitarianism asks what the outcomes of my actions are and describes how the proper course of action to be taken is the one that maximises the positive effect, which at its height is happiness. This was a popular theory in the 18th Century with philosophers noting that 'the greatest happiness of the greatest number that is the measure of right and wrong.'[1] The second theory is deontology which asks how I ought to act and is focussed on obligation and duty. This theory dictates that an act can be right, even if the outcome is bad. Immanual Kant believed that a person should 'act only according to that maxim through which you can at the same time will that it become a universal law.'[2] The theory outlines the requirement to follow the rules, such as International Humanitarian Law. However, unlike consequentialism it leaves little flexibility or ability to adapt to the situation. Nevertheless, it is deontology which feeds into the Just War tradition. This has three parts, the first at the political/strategic level deals with when to go to war and is known as *Jus ad Bellum*. The second is how to conduct a just peace and how to act post conflict and is termed *Jus post Bellum*, which is also mainly at the political and military strategic level. Finally, there is *Jus in Bello* which is concerned with the conduct of war. In the British Army this is covered by The Law of Armed Conflict (LOAC) and Rules of Engagement (ROE), both of which derive from International Humanitarian Law. *Jus in Bello* consists of four principles: military necessity – force should only be used with restraint, humanity – forces must honour human rights, proportionality – only the correct amount of force should be used and distinction – distinction between combatants and non-combatants must be respected. The final theory of normative ethics is virtue ethics in which character is the driving force for ethical behaviour. This theory dictates that people will naturally do what is good if they know what is right. This idea that ethics is about a person's character is attractive to most militaries who tend to have developed a system of virtue ethics. This chapter will first explore virtue ethics and the British Army Core Values which are based on them. It will then describe some of the problems of using just virtue ethics in combat situations, before broaching the use of the moral compass as a tool to assist with the use of virtue ethics in stressful situations. It will conclude with a case study investigating the use of the moral compass by Warrant Officer Thompson at My Lai.

---

1 'A Fragment on Government', Jeremy Bentham, 1776, https://plato.stanford.edu/entries/bentham/#:~:text=Bentham%20launched%20his%20career%20as,A%20Comment%20on%20the%20Commentaries.
2 'Groundwork of the Metaphysic of Morals', Immanual Kant, 1785, https://human.libretexts.org/Bookshelves/Philosophy/Ethics_(Fisher_and_Dimmock)/2%3A_Kantian_ethics.

Virtue Ethics

The theory of virtue ethics is concerned with the cultivation of an excellent human character who then conducts moral acts. Virtue ethics normally state that there are four cardinal virtues that underpin ethical actions, these are:

- *Prudence.* Prudence or wisdom is about understanding a situation as it actually is and not how you would wish it to be in order to identify the good in the situation. It involves the application of discernment, deliberation, decision and direction
- *Justice.* Justice is about fairness to each person
- *Fortitude.* Fortitude or courage is the will to face fear or endure harm in order to do good
- *Temperance.* Temperance or restraint is moderation based on self-knowledge, self-discipline and self-control.

The Ancients Greeks, such as Socrates and Aristotle, emphasised that virtue is practical, and that the purpose of ethics is to become good, not merely to know what is good. They believed that character came from habit and likened ethical character to a skill that can be acquired through practice. In Aristotelian philosophy the middle way, the mean, is the desirable meridian between two extremes, one of excess and the other deficiency. Therefore, there is the belief that 'The Golden Mean', the middle ground of moderation is the correct place to be, this middle ground and the extremes will be explored later in this chapter.

As most western militaries teach a system of virtue ethics they have tended to write about it in their doctrine. The British Army states that peoples 'values are specific beliefs that people have about what is important and unimportant, good and bad, right and wrong.'[3] It continues that leaders 'communicate the Values and Standards through everything that they do, both verbally and non-verbally, on and off duty.'[4] The US Army writes about how 'character consists of the moral and ethical qualities of an individual revealed through their decisions and actions.'[5] It also adds that 'adhering to the Army Values is essential to upholding high ethical standards of behavior' [sic].[6] Finally the Australian Defence Force describes how 'virtue ethics is

---

3 MOD, *The Army Leadership Code*, 6.
4 MOD, *Army Leadership Doctrine*, 5-1.
5 US Army, *ADP 6-22*, 1-14.
6 Ibid., 2-6.

distinct from other ethical theories in that it is not just concerned with a single act, but with how the repetition of an act, forming a habit, develops a person's character over time.'[7] These doctrines are mirrored by most western militaries which believe that the virtue, and therefore the character, of their personnel is the main pillar for the ethical performance of their organisations. However, as expressed in chapter 7, research has shown that the power of the situation is more dominant than character when it comes to peoples' behaviours in stressful situations. Nevertheless, it is important to understand how virtue ethics are understood in the context of the British Army.

The British Army's Core Values

The British Army's Core Values are based in the virtue ethics tradition. The difference between values and virtues will be explored later, but the British Army Core Values are the virtues of courage, discipline, respect for others, integrity, loyalty and selfless commitment. They are commonly referred to in the military by the acronym C-DRILS. A comprehension of what the British Army means about each value is:

- *Courage.* Courage is explained as recognising that 'soldiering has always demanded physical courage, to knowingly go into harm's way on behalf of the nation. Physical courage is required to risk life, take life, show restraint, endure hardships and focus on the task.'[8] The British Army also notes that equally important is moral courage which is defined as, 'the strength and confidence to do what is right, even when it may be unpopular and to insist on maintaining the highest standards of behaviour and decency. This earns respect and fosters trust.'[9]
- *Discipline.* Discipline is described as 'the primary antidote to fear and maintains operational effectiveness: it is supported by team loyalty, trust and professionalism. Discipline instils self-confidence and self-control. Good discipline means soldiers will do the right thing even under the most difficult of circumstances.'[10]
- *Respect for Others.* Respect for others is defined as including 'both those inside and outside of our organisation is not only a legal

---
7 ADF, *Military Ethics*, 16.
8 MOD, *The Army Leadership Code*, 8.
9 Ibid.
10 Ibid.

obligation, it is a fundamental principle of the freedom that our society enjoys'. ... 'We must treat everyone we encounter, as we would wish to be treated.'[11]

- *Integrity*. Integrity is explained as meaning individuals must be 'truthful and honest, which develops trust amongst individuals and welds them into robust and effective teams. Integrity is therefore critical to soldiering, as soldiers must have complete trust in one and other as their lives might ultimately depend on it.'[12]
- *Loyalty*. The doctrine states that loyalty 'binds all ranks of the Army together, creating cohesive teams that can achieve far more than the sum of their parts. The Nation, Army and Chain of Command rely on the continuing allegiance, commitment and support of all who serve.'[13] In also concludes that loyalty 'is not blind and must operate within the parameters of the other Values; it should not stop appropriate action to prevent transgressions by subordinates, peers or seniors.'[14]
- *Selfless-Commitment*. Finally selfless-commitment is explained as the requirement for 'the needs of the mission and the team [to] come before personnel interests. Ultimately, soldiers may be required to give their lives for their country, that is true selfless commitment.'[15]

In 2014, I conducted a British Army Core Values survey at the RMAS. There were 151 respondents from staff and Senior Term Officer Cadets. When asked how they considered the current Core Values '90% replied that they were either very satisfied or satisfied' with them and that 'they were relevant to modern officers.'[16] In the same survey respondents were asked if they considered the current British Army Core Values to be suitable and relevant. Two thirds of the respondents considered that the 'current values were well balanced and could not be added to.'[17] Of the third who did offer an additional value, the most popular was humility with '24 per cent recommending it, as

---

11 Ibid.
12 Ibid., 9.
13 Ibid.
14 Ibid.
15 Ibid.
16 Dennis Vincent, *Be, Know or do? An analysis of the Optimal Balance of the Be, Know, Do Leadership Framework in future Training at the Royal Military Academy Sandhurst.* (Sandhurst: Central Library, 2015), 8.
17 Ibid., 20.

an additional new value.'[18] In the RMAS 2023 research survey a question was asked to judge if the trainees understood the requirement to internalise the British Army Core Values and accept them as their own virtues. The question asked: It is OK not to follow the Army's Values when you are not in uniform and off duty? For the cohort with no previous military exposure the 'responses remained consistently biased strongly towards disagree' with between 97 to 99 per cent disagreeing.[19] For the end of training survey of all trainees the British response remained high with 99 per cent accepting the Army Core Values as their own. The researched commented that 'the very high and consistent acceptance of the Army values is not surprising given its prominence and emphasis at RMAS.'[20] However, the research also showed that 'only 80% of international Ocdts [Officer Cadets] appear to have internalised the Army values' as their own.[21]

There are some advantages to using virtue ethics. For example, 'people who have received a training on virtues might very well be less influenced by the situation they find themselves in than those who lack training.'[22] However as will be seen later in this chapter, the high figures of the RMAS 2023 research survey accepting the British Army's virtues as their own, does not automatically mean that these trainees applied these virtues in stressful situations, which highlights one of the main problems with virtue ethics. There are several other problems with using virtue ethics to regulate behaviour in stressful situations.

Problems with Virtue Ethics

Despite virtue ethics being the most popular ethical theory amongst western militaries, there are a number of problems with it as the basis for dealing with decision making in a stressful environment. These problems are dealing with complexity, reason, limited guidance, contamination, subjectivity and that they are culturally specific. Virtue ethics does not deal well with the type of complexity that leaders are faced with in the contemporary operating environment. In these volatile, uncertain, complex and ambiguous situations

---

18 Ibid.
19 Vincent and Muhl-Richardson, *An Analysis of the Effectiveness of the S-CALM Model of Ethical Leadership*, 20.
20 Ibid.
21 Ibid.
22 Olsthoorn, 'Situations and Dispositions: How to Rescue the Military Virtues from Social Psychology', 87.

there are often moral dilemmas where there is more than one virtuous thing to do. The US Army summarises this issue when it states that 'to be an ethical leader requires more than merely knowing the Army Values. Leaders must be able to live by them to find moral solutions to diverse problems.'[23] The second problem is the issue of reason. Virtue ethics does not allow individual choice or reason when making decisions. When commenting on the British Army's Values and Standards pamphlet, ethicist Paul Robinson, in *Ethics Training and Development in the Military* noted 'the pamphlet follows a "virtue ethics" approach, detailing the virtues required of the soldier, and implying that the way to ensure proper ethical behavior [sic] is through the inculcation of the necessary virtues.'[24] The third problem is limited guidance. Although virtue ethics praises various character traits it provides insufficient guidance on how to make decisions when confronted with the need to solve moral dilemmas. Paul Robinson again comments that 'teaching soldiers that they must be brave, loyal, and so forth, does not tell them what to do when there are conflicts between the requirements of various virtues.'[25] Contamination is the next problem. As has already been identified, even people with good character can become contaminated by the power of the situation. Therefore, a virtuous character alone is not enough. In some situations what might seem to be the right action, is unethical. For example, in the Baha Musa murder inquiry, the investigation team faced what was termed a 'green wall of silence' by the members of the 1st Battalion the Queen's Lancaster Regiment.[26] The soldiers wanted to demonstrate their value of loyalty to their unit, no soldier had the moral courage to break from the group and tell the truth about the abuse that had taken place. The penultimate problem is subjectivity. The idea that everyone operates most of the time within 'The Golden Mean' is ambiguous. Every individual's mean is relative and what might appear moderate to one could be extreme to another. How do you judge loyalty? How do you judge courage? These questions are difficult to answer as they are so subjective, everyone is approaching virtue ethics from their own perspective. It is only when a person's actions are at the extreme that they can be clearly judged by most observers to have objectively broken a virtue. In 2015 the Reverend Philip McCormack was a British Army padre

---

23 US Army, *Army Leadership*, 2-7.
24 Paul Robinson, 'Ethics Training and Development in the Military', *Parameters*, Vol. 37, No. 1 (2007), 29.
25 Ibid., 31.
26 MOD, The Aitken Report: An Investigation into Cases of Deliberate Abuse and Unlawful Killing in Iraq in 2003 and 2004 (MOD, 2008), 24.

Objective	Subjective	Mean	Subjective	Objective
Cowardice		Courage		Recklessness
Disorder Neglect		Discipline		Pernickety Over critical
Ignorance Disrespect		Respect for others		Unthinking Uncritical Acceptance
Deceit		Integrity		Saintliness
Treachery		Loyalty		Blind obedience
Selfishness		Selfless commitment		Martyr

Table 11.1 British Army Core Values Spectrum

who tried to codify this idea of objectively judging the British Army Core Values by displaying them on an ethical spectrum. In this spectrum 'The Golden Mean' was green and in the centre, with subjective deviance in amber and the extreme objective behaviour in red. A diagrammatic representation of Philip McCormack's ideas is shown in Table 11.1.

The final problem with virtue ethics is that they are culturally specific. Different cultural groups have conflicting opinions on what constitutes an ethical virtue. For example, the US Army has a set of values which are, loyalty, duty, respect, selfless service, honor [sic], integrity, and personal courage. These are similar to the British Army, but they do not have the value of self-discipline, whilst having the additional values of duty and honour. Paul Robinson explains how virtues are 'desirable characteristics of individuals, such as courage', while values correspond to 'the ideals that the community cherishes, such as freedom.'[27] The work by the German Army in this area is very interesting as it highlights their understanding of this difference and emphasises cultural differences. The German Army believes that 'values describe what you as a soldier or civilian employee of the Bundeswehr support, even with your life. Virtues describe how you fulfil your tasks and duties.'[28] Therefore their values reflect the overarching ethical ethos of the organisation and include 'freedom, human dignity, democracy, peace, justice,

---

27 Paul Robinson, 'Introduction: Ethics Education in the Military' in *Ethics Education in the Military*, eds P. Robinson, N. de Lee, and D. Carrick (Aldershot: Ashgate, 2008), 5.
28 Bundeswehr, *Making the Right Decision – Acting Responsibly*. (Bundeswehr, 2013).

British Army Values	German Army Virtues
Courage	Brave
Self Discipline	Discipline
Respect for Others	Fair, tolerant and open to other cultures. Comradely and considerate
Integrity	Truthful
Loyalty	Loyal and conscientious
Selfless commitment	
	Competent and cable of learning
	Able to distinguish right from wrong

Table 11.2 British Army Values and German Army Virtues

equality and solidarity.'[29] Their virtues are more about individual morals and in Table 11.2 can be compared to the British Army Core Values.

As can be seen, the German Army does not have an equivalent for selfless commitment but does include two other virtues of being competent and able to tell right from wrong. A comparison against other militaries would find the same issue of each having a slightly different culturally driven concept about what virtues were required.

Moral Compass

Some may question why this is a moral and not an ethical compass? What is the difference between the two words? The philosopher Professor A. C. Grayling suggested that 'ethics is to do with questions of right and wrong, but it includes more culturally-specific interests and attitudes and is more about character and virtues, whereas morality is not hypothetical and is more about fixed rules concerning what is right and wrong.'[30] In simple terms morality is about personal norms and is defined as 'relating to the standards of good or bad behaviour, fairness, honesty, etc. that each person believes in.'[31] On the other hand, ethics is about group consensus and is defined as 'relating to

---

29 Ibid.
30 Professor A C Grayling lecture: ethics versus morality, Warminster School, 2015, https://www.warminsterschool.org.uk/professor-a-c-grayling-lecture-ethics-versus-morality/
31 Cambridge Dictionary, https://dictionary.cambridge.org/dictionary/english.

beliefs about what is morally right and wrong.'[32] The concept of the moral compass has been defined as being 'a natural feeling that makes people know what is right and wrong and how they should behave.'[33] In the virtue ethics tradition character is the driving force for ethical behaviour, but the moral compass combines some elements of all three normative ethics theories. From consequentialism comes the idea that a morally right action produces a good outcome and from deontology comes the idea that acting in a moral way is an obligation or duty. Mitchell Hall in *Why Leaders need a Morality Check* stated that 'one of the most important things a junior officer must remember is that there is an ethical true north towards which the officer can align a moral compass; "right is right", no matter who says otherwise.'[34] The objective of having a moral compass is a *zeitgeist* that points true and does not swing from good to bad. It requires the ability to display moral courage in difficult circumstances and is vital for ethical leaders. Developing a moral compass must be done over a period of time and during everyday activities. Phillip McCormack noted that 'the cultivation of doing the 'right' in the everyday and mundane shapes behaviour and begins to develop the character required to face and overcome the ethically wicked problems potential enemies will seek to create.'[35] The continual habit of doing the right thing allows leaders to prepare for the stress brought on by the power of the situation. Mitchell Hall asserted that 'if combat leaders have their moral compasses set before the fighting starts, they can prevent tragedies', he also perceived that 'if a soldier, sailor, or marine enters combat without taking time to consider where they stand morally and ethically, it is already too late.'[36] Therefore when on operations a leader must be prepared to demonstrate moral courage to continue to keep their moral compass pointing true. Although military doctrines do not mention the moral compass much, they do discuss the need for moral courage. The British Army state that 'Moral Courage is the characteristic that all of the other Values and Standards depend', the US Army have a similar message stating that 'Moral courage is the willingness to stand firm on values, principles, and convictions.'[37] The Australian Defence Force

---

32 Ibid.
33 Ibid.
34 Mitchell Hall, 'Why Leaders need a Morality Check', *United States Navel Institute*, Vol. 132 No. 4 (2006), 69.
35 Phillip McCormack, Preparing Professional Military Forces to Face Ethical Challenges in Future Military Operations, EuroISME Presentation, 2015, 14.
36 Hall, 'Why Leaders need a Morality Check', 68.
37 MOD, *Values & Standards of The British Army*, 11. US Army, *ADP 6-22*, 2-5.

believe a leader must 'exhibit moral courage and create the conditions so that the rest of your team can demonstrate it as well.'[38] The Australians' also consider a leader 'requires the moral courage to question group processes; call out transgressions, change practices and behaviour; hold people to account; and recalibrate the group back to the expected behaviours of our profession of arms.'[39]

As operations continue and ethical drift begins to take effect it can act like a magnet that can pull a leader's moral compass off true. As mentioned earlier, a leader must not allow their moral compass to swing wildly from good to bad. Patrick Bury recalled that as a platoon commander he tried to 'create a culture whereby you are making the right decisions almost always.'[40] Therefore, leaders need consistency in their actions, the predictability outlined in chapter 9. They cannot enforce the rules one day and break them the next, this inconsistency will quickly lead to a lack of trust. The implication is that people 'tend to make the easy, rather than the ethically correct decision under the strain of combat.'[41] This erosion of the ability of ethical drift to dictate decision making can be seen in microcosm in the results of the RMAS 2023 research survey. In the survey, trainees were asked two questions to judge their understanding of their moral compass in difficult situations. The first question asked: It is justifiable to kill a civilian in order to achieve the mission? Whilst a second question in a different area of the questionnaire asked: It is justifiable to kill a civilian in order to save a soldier? For the cohort with no previous military experience the research showed in both questions that there 'was a shift in responses here over time, between the first and final timepoints there was a statistically significant decrease in the average response.'[42] For all trainees, the responses to both questions during the stress of the Stabilisation Exercise demonstrated some ethical drift on what might have been expected. For the first question the British trainees answered a split of '31% agreed, 21% were not sure and 48% disagreed.' This showed nearly a third were prepared to kill a civilian, whilst almost a half were not. For the international trainees only '7% agreed, 13% were not sure and 80% disagreed.'[43] International

---

38 ADF, *Leadership*, 13.
39 ADF, *Military Ethics*, 35.
40 Bury, 'Maintaining Morality at the Tactical Level', 121.
41 Dennis Vincent, 'Towards Jus Post Bellum: Ethical Warfare for Stabilisation in Iraq and Afghanistan' in *Jus Post Bellum: Restraint, Stabilisation and Peace*, ed Patrick Mileham. (Leiden: Brill Nijhoff, 2020), 360.
42 Vincent and Muhl-Richardson, *An Analysis of the Effectiveness of the S-CALM Model of Ethical Leadership*, 19.
43 Ibid.

Ocdts [Officer Cadets] demonstrated a stronger moral compass and more consideration than their UK counterparts. These results seemed to demonstrate the mission focus mindset of the tired British trainees, whilst there was more compassion amongst their international counterparts. For the second question, both groups were nearer in their responses with British trainees recording '30% agreed, 30% were not sure and 40% disagreed. For international Ocdts, 20% agreed, 40% were not sure and 40% disagreed.' The research noted that the 'lower than expected level of disagree responses here suggests that there is a broad lack of understanding about the demands of sacrifice in this cohort.'[44] The research concluded that the 'attitude towards civilian lives became more evident in the final term and more use of the RMAS 'Ethical Decision Making Coaching Tool' on Senior Term exercises might assist in changing this trend.'[45] Therefore even in a highly monitored environment unless the ethical education is reenforced and personnel are held to account on field training and operational deployments, there will be ethical drift.

The Moral Compass Test

It can be easy to do the right thing when you are being held to account. This is often termed the Hawthorne effect, which derives its name from a series of experiments carried out in the Hawthorne Works of the Western Electric Company, Chicago, between 1927 and 1933. The Hawthorne effect is defined as 'the tendency for people to behave differently when they know they are being studied.'[46] However, Thomas Macaulay speaking in the 19th Century said that 'the measure of a man's real character is what he would do if he knew he would never be found out.'[47] This concept that a person should conduct themselves in a way that is ethical even if they are not being watched or their actions would not be uncovered form the basis of the moral compass test. British Army personnel are required to act virtuously in any circumstance that they might face, whatever the external stimuli, and do the morally right thing. This translates to the moral compass test which suggests that a person should 'do the right thing on a bad day when no one is looking'. This test has been turned into a mantra at the Royal

---

44 Ibid.
45 Ibid., 21.
46 D. Davis and V. Shackleton, *Psychology and Work*. (London: Methuen, 1975), 55.
47 Forbes Quotes, https://www.forbes.com/quotes/author/thomas-b-macaulay/.

Military Academy Sandhurst, where trainee officers are taught to learn it and put it into practice in all circumstances. The objective being that this becomes the new cognitive norm and their intuitive way of acting. General Sir James Glover, when Commander-in-Chief UK Land Forces, wrote of the internalising of this way of thinking that 'it defines a line he dare not cross and deeds he dare not commit, regardless of orders, because those very deeds would destroy something in him which he values more than life itself.'[48] The moral compass test causes an individual to ask the question 'am I doing the morally right thing? Depending on the answer they will either continue or change their behaviours.

Summary

A leader therefore needs to prepare themselves and their team to act in an ethical way under the power of the situation. By applying the moral compass to routine activities, the leader and their team can build up resilience. Ethical leaders can also prepare themselves and their teams for the stress of conflict by discussing case studies and by the practice of ethical reactions to scenarios. The more this is done, the more likely ethical actions will become the default setting when under pressure to act under stress. Some people may already have a strong moral compass and a tendency to act for good in a stressful situation. Leaders need to consider this 'Banality of Good' in every one of their followers and ensure that they tap into this and allow moral actions to flow. Finally, the third key question in a stressful situation a leader needs to ask is – Am I doing the morally right thing?

Moral Compass Case Study – Warrant Officer Thompson at the My Lai Massacre

The My Lai massacre has been mentioned many times, but this case study identifies the actions of the real hero who stopped the killing and brought the perpetrators to justice; Warrant Officer (WO) Hugh Thompson. WO Thompson was tasked to pilot a reconnaissance helicopter with a crew of Specialists Glenn Andreotta and Lawrence Colburn during the operation. At about 0900 hours WO Thompson noticed some wounded civilians in a ditch and marked them with green smoke, to signal medical assistance was

---

[48] James Glover, 'A Solider and His Conscience', *Parameters*, Vol. 13, No. 1 (1983), 56.

required. After refuelling he flew over the ditch again and noticed that the wounded had been killed and that there were more bodies in the ditch. At this point, he sent the radio message about the number of bodies which ended 'there's something wrong here.'[49]

WO Thompson had begun to understand that the actions of C Company were not ethical. Then Captain Ernest Madina approached another female civilian casualty that WO Thompson had marked with green smoke, he kicked and then shot her. WO Thompson's moral compass was now at odds with the actions on the ground and when he saw a group of civilians being herded into a bunker, he knew that he had to take action. He landed the helicopter between the civilians and the US troops and told Specialists Colburn and Andreotta to shoot the US troops if they opened fire on the civilians, saying 'If these bastards open up on me or these people, you open up on them. Promise me!'[50] He confronted the platoon commander Second Lieutenant Stephen Brooks saying 'Hey listen, hold your fire. I'm going to try to get these people out of this bunker. Just hold your men here.' Brooks replied 'Yeah, we can help you get 'em out of that bunker—with a hand grenade!'[51] WO Thompson, despite being of a junior rank took control and extracted 11 civilians from the bunker. He then called in two Huey gunships to extract them from the scene. He reported to the brigade headquarters that there was a massacre occurring at My Lai and that the attack should be stopped immediately.

Such was WO Thompson's moral courage that having made the decision to intervene, he also agreed to give evidence at the courts martials that were held in 1970. This made WO Thompson exceptionally unpopular, and he was ostracized by much of the Army. He also recalled that there were 'death threats at three o'clock in the morning, mutilated animals on your doorstep, and I'm sitting here just as confused as hell. What in the world is wrong? What we did was right. What we did had to be done.'[52] After the Vietnam War there was a realisation that rather than being rejected for the actions he took; he should be held up as an example. On 6 March 1998, exactly 30 years after the actions he was awarded the Soldier's Medal along with Specialists Colburn and Andreotta (posthumously). The Soldier's Medal was awarded

---

49 Warrant Officer Historical Foundation, The Forgotten Hero of My Lai: The Hugh Thompson Story, https://warrantofficerhistory.org/PDF/Forgotten_Hero_of_My_Lai-WO_Hugh_Thompson.pdf.
50 Ibid.
51 Ibid.
52 Thompson, *Moral Courage in Combat*, 12.

for their life-saving actions at the village of My Lai and is the US Army's highest award for bravery not involving direct contact with the enemy. At the ceremony, Major General Michael Ackerman said of the helicopter crew that 'it was the ability to do the right thing even at the risk of their personal safety that guided these soldiers to do what they did.'[53] After this award WO Thompson, now a retired major, gave presentations on professional ethics at all US military academies. The actions of WO Thompson at My Lai demonstrate that if a person has a good moral compass and the answer to the question 'is this the morally right thing?' is no; they must take action despite what the personal consequences might follow.

---

[53] Warrant Officer Historical Foundation, The Forgotten Hero of My Lai: The Hugh Thompson Story.

12

Application of the S-CALM Model

Introduction

In early 2003 as Battalion Second in Command, I was selected to join the US Army Corps HQ planning team that were preparing for the invasion of Iraq. I strongly believed that the invasion of Iraq was an unjust war and went against my principles as a Christian. However, I had been in the British Army for 18 years and this was an opportunity to go and fight in a war. I really struggled with what to do, my instinct was to deploy to the fight, but my accountability and moral compass would not let me take part in what I perceived to be an unjust war. I decided that my only option must be to resign my commission. I duly submitted my resignation letter to the Commanding Officer. A few days later, I was told that the American's no longer wanted a British planner and I was stood down from the deployment. I was both pleased and disappointed, as this decision had cost my career. However, immediately after the deployment was cancelled, the Commanding Officer came into my office and ripped up my resignation letter. Shortly after, the battalion deployed to Northern Ireland and on return, quickly redeployed to Afghanistan. I was then posted from the battalion and found myself appointed as the Executive Officer to the UK Counter Insurgency Short Term Training Team in Kuwait and Iraq. This deployment to Iraq I could reconcile as I felt that the UK was accountable for the damage that the invasion had created, and I was going some way to put things right. My moral compass was also at ease with this deployment as it was for the good of the Iraqi people. During this deployment, which covered the changeover of the US Army Corps in Iraq, I taught over 5,000 US commanders how to conduct ethical counter insurgency operations. My decision to resign was without doubt an exceptionally difficult one. However, I believed that I had to take heed of my moral compass and do what I considered to be the right ethical action at that time, regardless of the potential personal cost.

The application of the S-CALM model is very similar to the flow of unethical events as described in chapter 5. Leaders will find themselves in a stressful situation with situational influencers and common behaviours, but they are now looking to recognise them and mitigate them by using their knowledge of accountability, leadership and the moral compass. This chapter explains how this application of the model unfolds. It will also expand on the STOP Protocol, which is the simple acronym for taking positive action in stressful situations. It will explain how this led into the development of the Ethical Decision Making Coaching Tool, which can be used to assist in the training of leaders. It will also outline the details of the RMAS ethical leadership research survey, which was conducted in 2023. Finally, it will conclude with a summary which includes the overarching Ethical Hierarchy.

Application

When under pressure the first thing that a leader should identify is that they are in a stressful or toxic situation which is having power over them and their team. Once they have faced up to this, the next step is to recognise any situational influencers that might be increasing the power of the situation on them. Some may have been identified and dealt with before the situation began, such as a lack of resource for example. However once in the situation, leaders should try and identify influences affecting both themselves and their team. Any situational influencers recognised should be mitigated by the actions already discussed. The leader also needs to be aware of the common behaviours that they themselves and their team are susceptible to before they get into stressful situations. This should assist leaders and teams from being susceptible to these common behaviours, or at least being on the lookout for their signs and symptoms. Nevertheless, commanders should be ready to mitigate any of the common behaviours that they might recognise happening as the situation develops. A diagrammatic representation of the application of the S-CALM model is at Figure 12.1.

Once a leader has recognised the situational influencers and common behaviours, they need to apply their accountability by asking the key question are my actions lawful, appropriate and professional? They then

Figure 12.1   Application of the S-CALM Model

need to consider their leadership and judge if it is time to apply the loneliness of command? Finally, they are required to reflect on their moral compass and ask the final question am I doing the morally right thing? These key questions are the final part of the STOP Protocol, which will be explained later, but first the application of the model requires both reflection and self-awareness.

Reflection and Self-Awareness

As described in chapters 5 and 7, the application of the model requires reflection using System 2 thinking and display by the leader of emotional intelligence, especially self-awareness and self-regulation. It has been identified that 'self-reflection is central to leadership. The more you self-reflect the better you know yourself: your strengths, weaknesses, abilities and areas for development.'[1] By reviewing past experiences and considering future situations, a leader will be able to prepare themselves and their teams better for the stress of a situation. The exercise at the end of chapter 8, which asked what you considered you and your teams default common behaviours when not under stress, was aimed at conducting this reflection. It is better to have considered issues before they happen and prepare than regret them after. A good practice for this is the pre-mortem, in which 'participants work on the assumption that the strategy has failed and so question what went wrong.'[2] Pre-mortems do however require a team that is mature and able to consider poor outcomes and failure without turning them into reality. Suzy Welch, a well known US business book writer, wrote in *10-10-10: A Life-Transforming Idea* that people should consider 'what are the consequences of my decision in 10 minutes? In 10 months? And in 10 years?'[3] This idea of contemplating

---

1   Kemp, *British Justice, War Crimes and Human Rights Violations*, 14.
2   MOD, *The Good Operation*, 91.
3   The Rule of 10-10-10, Suzy Welch, https://www.oprah.com/spirit/suzy-welchs-rule-of-10-10-10-decision-making-guide/all.

both the short- and long-term effects of a decision should allow a leader to put their judgement into perspective. Leaders must therefore make time to reflect on past experience and identify lesson learned in them. They must also view possible future scenarios and consider what common behaviours they and their teams might display.

Research has shown that 'self-aware leaders understand who they are and what they want to achieve.'[4] The requirement for self-awareness, that internal state where a leader understands their strengths and weaknesses, their own resources to be resilient and the impact they have on their team is important. It has been assessed that 'an inner compass of self-awareness' allows a leader 'to walk the tight rope of leadership.'[5] The British Army describe how a leader 'requires self-awareness and understanding of how their actions and behaviours impact those they lead and how they are perceived by those around them. Self-awareness is the bedrock from which all other levels of leadership are built.'[6] The US Army also deems that 'self-aware leaders are more responsive to situational and interpersonal cues regarding actions to take.'[7] Finally the Australian Defence Force believes that both 'self-awareness and self-reflection are critical skills required for ethical decision-making.'[8] In addition to reflection and self-awareness there is a necessity to be able to self-regulate. Self-regulation is about managing the internal state to ensure that impulses, feelings and emotions are kept under control. This act of taking control is best achieved when people are able to engage their System 2 thinking. Daniel Kahneman in Thinking, Fast and Slow states that 'the operations of System 2 are often associated with the subjective experience of agency, choice, and concentration.'[9] Once they are in a stressful situation the leader will need to use their System 2 thinking whenever there is time to do so, to recognise how the situation may be driving their thought processes. This may be as simple as stopping and taking a deep breath, conducting an estimate or using participative leadership and asking for opinions on the situation. To assist with this there is the STOP Protocol.

---

[4] Mendemu Showry and K. V. L. Manasa, 'Self-Awareness-Key to Effective Leadership', *IUP Journal of Soft Skills*, Vol. 8, No. 1 (2014), 23.
[5] Ibid., 15.
[6] MOD, *Army Leadership*, 1-9.
[7] US Army, *ADP 6-22*, 6-3.
[8] ADF, *Military Ethics*, 37.
[9] Kahneman, *Thinking, Fast and Slow*, 22.

STOP Protocol

The STOP Protocol is a simple acronym that can be used by leaders to remember how to apply the S-CALM model under the stress of the situation. Each letter of STOP stands for a step which is conducted to allow for a leader not to be driven into unethical actions.

S – Stop what you are doing. The first step is to simply stop doing whatever it is that might be perceived as not right. If a leader believes that they or their team are acting in an unethical way they should call a stop to it. This operates as a brake on the spiral of violence in which the dominance of System 1 thinking and the social pressure to be consistent with our actions, allow small deeds to develop into bigger ones which can get out of control. The ability to be proactive, take the initiative and give a clear order to cease an activity is the vital first step in the prevention of further unethical actions.

T – Take a few deep breaths. Research has demonstrated that 'people can behave in an ethical way—they just need time' and taking a few deep breaths is a way of gaining that time.[10] In many of the case studies investigated in this book, soldiers have not taken that deep breath until after they have committed their transgression because they allowed their System 1 thinking to drive them to unconsidered actions. Daniel Kahneman in Thinking, Fast and Slow describes how 'the way to block errors that originate in System 1 is simple in principle: recognize the signs that you are in a cognitive minefield, slow down, and ask for reinforcement from System 2.'[11] British Army doctrine states that 'finding time to think can be difficult, but leaders must find opportunity and space to do so in order to make sound judgements.'[12] Taking a few conscious breaths allows a leader to engage their System 2 thinking and to move from an emotional to a rational state. A good system to employ is what is termed Combat Tactical Breathing which is practiced by the US Navy Seals. Combat Tactical Breathing 'focuses on slowing the breathing rate down by breathing through the nostrils, counting to four for each inhale and exhale' and was explained in chapter 6.[13] This simple second step of taking a few deep breaths allows the leader to move from a state of

---

10 Shaul Shalvi, Ori Eldar and Yoella Bereby-Meyer, 'Honesty Requires Time (and Lack of Justifications)', *Psychological Science 2012*, Vol. 23, No. 1264 (2012), 1269.
11 Kahneman, *Thinking, Fast and Slow*, 407.
12 MOD, *Army Leadership Doctrine*, 3-7.
13 'How To De-Stress In 5 Minutes Or Less, According To A Navy SEAL', Forbes, https://www.forbes.com/sites/nomanazish/2019/05/30/how-to-de-stress-in-5-minutes-or-less-according-to-a-navy-seal/?sh=7705ec7b3046.

being under the power of the situation to one in which they gain control of their cognitive abilities.

O – Observe any situational influencers and common behaviours. Once rational thinking has been allowed to work, the leader can observe any situational influencers that are causing them to behave in ways they would not wish to. For example, is there a hostile environment or are emotions running high? They can also identify any common behaviours, which as earlier identified might be ones that they normally default to on a daily basis or could be ones specific to the situation. For example, are they conforming to the group, or are they displaying cognitive dissonance? This step allows a leader to gain clarity on how they and their team may be affected by the power of the situation and leads on to the final important step.

P – Proceed considering accountability, leadership and moral compass. In many situations that a leader will find themselves although they may have identified unethical actions, they may need to continue with the task regardless of this as they must complete their mission. Therefore, they will need to proceed, but in a different way. They should now consider their accountability, leadership and moral compass to ensure that they continue with their mission in an ethical manner. To assist in this, they should ask themselves the three key questions which were described earlier; they are:

- Accountability: Are my actions Lawful, Acceptable and Professional (LAP)?
- Leadership: Do I need to apply the loneliness of command?
- Moral Compass: Am I doing the morally right thing?

The STOP Protocol is an important tool for a leader to have in the complex situations they will find themselves in from a difficult day in the office to combat. Although there are other theoretical ideas, such as the Just War theory, research has shown that 'an unbridgeable divide exists between war as it is and war as the moralist would like it to be or imagines it to be. The deliberations of moral philosophers, including just war theorists, bear little relation to the harsh realities of war.'[14] Therefore this simple applied tool allows a leader under stress to guide themselves and their team into ethical action.

---

14 Coates, 'Culture, the Enemy and Moral Restraint in War', 208.

Ethical Decision Making Coaching Tool

To assist in the practical application of the S-CALM model on field training exercises, another tool was developed to assist in the coaching of those who had faced an ethical dilemma. This tool was devised in conjunction with Major Ben Ordiway, a military instructor at the United States Academy West Point. Major Ordiway had been working on what was termed a Moral Terrain Coaching Tool which allowed for a coach to conduct reflection with a person who had made a decision of a moral nature. Major Ordiway's belief was that there needed to be more emphasis on assisting with moral decisions, rather than just a focus on command decisions. Research concurred that 'military ethics training that incorporates awareness of the importance of emotion in moral judgment can help warriors under stress rapidly and non-consciously identify and assess the weight of diverse morally salient considerations.'[15] Major Ordiway visited the Royal Military Academy Sandhurst in 2023 and trialled the US Army tool. After further collaboration a British version of the tool which is based on the S-CALM model was developed and called the Ethical Decision Making Coaching Tool. This tool is used on Royal Military Academy Sandhurst final term exercises where there is a degree of complexity in the scenarios and the possibility that unethical orders or actions might be conducted. The tool is simple to use and takes around 30 minutes to complete. When it has been used, the trainees have gained a much better understanding of ethical decision making. A copy of the tool along with the guidance for its use is at Appendix 2.

Ethical Leadership RMAS Research Survey

Extracts from the Sandhurst Occasional Paper An Analysis of the Effectiveness of the S-CALM Model of Ethical Leadership at the Royal Military Academy Sandhurst have been used extensively throughout the second half of this book to illustrate the research on the model. This paper came from a research survey which was conducted with the Regular Commissioning Course 231 cohort, focused on the internalisation of the S-CALM model of ethical leadership during their year of training at RMAS. The officer trainees of Regular Commissioning Course 231 completed the same survey three times during their training at RMAS. Initially on arrival in week one before any

---

15 Mitt Regan and Kevin Mullaney, 'Emotion, Ethics, and Military Virtues', *Journal of Military Ethics*, Vol. 22, No. 3-4 (2023), 269.

ethical education. Then on week four of the second term, after the completion of their CABS ethical leadership education and finally during Exercise Templars Triumph, a complex stabilisation exercise, in week seven of the final term. The method used for this research was a short online survey. The survey was called a 'Decision Questionnaire', so as not to give an idea that it was about ethics. The survey questionnaire asked a series of demographic questions and then 15 questions which used a seven-point likert scale, to allow the officer trainees to give nuanced answers.

The research was designed to investigate two research questions. The first question was aimed at gaining an understanding of the effectiveness of ethical leadership education at RMAS. The second question was intended to understand how effective the S-CALM model was at guiding a trainee officer's moral judgement. Research question one examined the answers given to the survey by those who had arrived at Sandhurst with no previous military training. This caveat was put in place as it was assessed that those who had been in the Army Cadet Force, University Officer Training Corps, Reserves or Regular Army would have already been exposed to the basics of military ethics and the Army's Values and Standards. Therefore, the cohort for this part of the research was restricted to UK and International officer trainees who reported no previous military experience. The results from all three tranches of surveys were assessed to identify changes in responses and to understand the effectiveness of ethical leadership education at RMAS for this cohort. Research question two examined the answers given by all respondents in the third survey which was conducted during Exercise Templars Triumph in week seven of their final term. This exercise was selected as the officer trainees were tired and not focused on ethical thinking. Results were assessed for this final survey from a simple point of view of all respondents, but further in-depth analysis was also considered using the variables of age, experience, gender, education and nationality identify to investigate differences in response. The results of this analysis assisted in the understanding of how effective the S-CALM model is at guiding officer trainee moral judgement. The research summarised its finding as follows:

> The outcome of this research suggests that the Situational Influencers were well understood by the cohort. As for the Common Behaviours, it appears that these are also reasonably understood, but that additional focus on othering and obedience is required. In the Leadership area, there is a requirement for a better understanding

of the requirement for an 'air gap' between leaders and their followers in order to make the difficult decisions when required.[16]

If you completed the questionnaire in Appendix 1 as suggested at the end of chapter 1, you may wish to complete it again at Appendix 3 and compare your answers to see if reading this book has assisted in your understanding of how to deal with some of the dilemmas presented.

Summary – The Ethical Hierarchy

Most people across the world hold to the ideas of an ethical foundation, which respects life, liberty and justice. In addition, most would also have ethical principles, including the concepts of equality, worth and dignity. Nevertheless, when people join the military, they are expected to accept the virtues of the organisation on top of these shared foundations and principles. For example, in the British Army this takes the form of agreeing to a set of values and standards which are not common to those they might have held whilst in civilian society. These foundations and 'principles of the right to life and liberty reflect some of the obligations on officers and soldiers detailed in the Geneva Conventions and which are part of the broader Laws of Armed Conflict.'[17] However, as leaders there needs to be more substance to the understanding of how to apply ethical leadership under stress and when exposed to the power of the situation. Ethical leadership has been expressed as the active application of ethical thinking. It is conducted by those who recognise unethical behaviours and take steps to correct them. This extra requirement of a leader is encapsulated in the S-CALM model, which allows leaders to make sound judgements and decisions when faced with difficult ethical dilemmas in stressful situations. These notions of an Ethical Hierarchy, from foundation, principles, organisational virtues to the active application of ethical leadership are captured in Figure 12.2.

A major part of being an ethical leader is to reflect and prepare for situations before being exposed to them. Having this 'knowledge that situational forces determine most of our conduct' is important as it allows leaders to anticipate what might happen and to mitigate their actions.[18]

---

16 Vincent and Muhl-Richardson, *An Analysis of the Effectiveness of the S-CALM Model of Ethical Leadership*, 21.
17 MOD, *Values and Standards*, 4.
18 Vincent, 'Towards Jus Post Bellum', 80.

The S-CALM Model

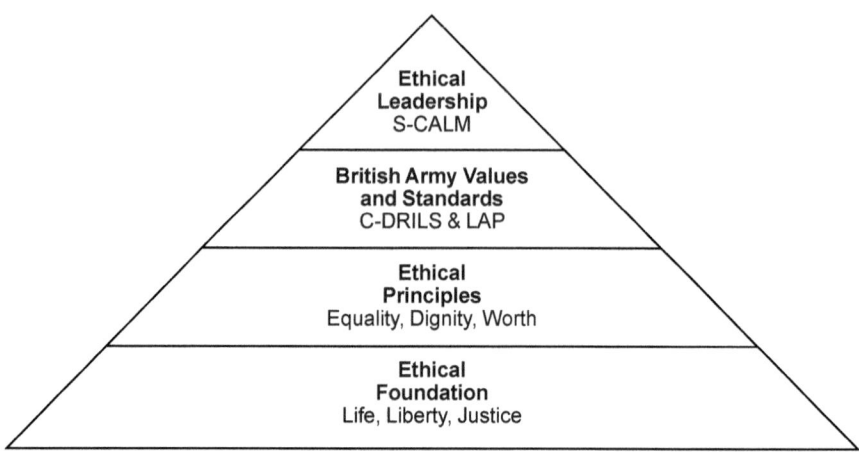

Figure 12.2   **Ethical Hierarchy**

However, it is important that ethical leaders not only prepare themselves, but also educate their teams in the moral component of fighting power. This education needs to be more than just teaching on virtue ethics but should also 'aim at giving insight into the situational forces that make unethical conduct more likely to take place.'[19] The aim should be to develop happy, well-motivated, well-educated and empowered teams, as these are less likely to commit unethical behaviours. It is important to harmonise leadership, behavioural science, military ethics and ethical leadership and not treat them as separate subjects. Ethical leadership situations should be incorporated into all training to reduce the risk of unethical actions taking place, especially on operations. The S-CALM model offers an excellent way for leaders to achieve ethical outcomes. Understanding the situational influencers that enhance the power of the situation and what common behaviours are likely to occur are important. Holding themselves and their teams to account, displaying strong leadership and ensuring their moral compass is true is also vital. The STOP Protocol of stopping, taking a breath, observing the situational influencers and common behaviours and proceeding in an ethical way provides a useful handrail. As does the asking of the three vital questions: Are my actions Lawful, Acceptable and Professional? Do I need to apply the loneliness of command? and Am I doing the morally right thing? The S-CALM model gives a leader a simple way to apply ethical decision making in the volatile,

---

19 Olsthoorn, 'Situations and Dispositions: How to Rescue the Military Virtues from Social Psychology', 91.

uncertain, complex and ambiguous environment of the 21st Century. It can be used under the stress of a bad day in the office or in combat and should be practiced as much as possible, so it becomes an instinctive tool for all leaders.

APPENDIX 1: DECISION MAKING QUESTIONNAIRE 1

Please tick or circle the box which best reflects your answer to the 15 questions below:

Decision Questions

1. It is OK for a platoon commander to use the first names of their soldiers?

Strongly agree	Agree	Slightly agree	Neither agree nor disagree	Slightly disagree	Disagree	Strongly disagree

2. You can break the rules of war if you believe that they are not useful for your mission?

Strongly agree	Agree	Slightly agree	Neither agree nor disagree	Slightly disagree	Disagree	Strongly disagree

3. You should always obey an order by a senior commander?

Strongly agree	Agree	Slightly agree	Neither agree nor disagree	Slightly disagree	Disagree	Strongly disagree

4. A commander should never sleep when their soldiers are awake?

Strongly agree	Agree	Slightly agree	Neither agree nor disagree	Slightly disagree	Disagree	Strongly disagree

Appendix 1: Decision Making Questionnaire 1

5. It is OK not to follow the Army's Values when you are not in uniform and off duty?

Strongly agree	Agree	Slightly agree	Neither agree nor disagree	Slightly disagree	Disagree	Strongly disagree

6. It is justifiable to kill a civilian in order to achieve the mission?

Strongly agree	Agree	Slightly agree	Neither agree nor disagree	Slightly disagree	Disagree	Strongly disagree

7. It is always best to go with your initial intuitive decision?

Strongly agree	Agree	Slightly agree	Neither agree nor disagree	Slightly disagree	Disagree	Strongly disagree

8. An enhanced emotional state leads to an increase in unethical behaviour?

Strongly agree	Agree	Slightly agree	Neither agree nor disagree	Slightly disagree	Disagree	Strongly disagree

9. It is justifiable to kill a civilian in order to save a soldier?

Strongly agree	Agree	Slightly agree	Neither agree nor disagree	Slightly disagree	Disagree	Strongly disagree

10. Fatigue has little affect the way that commanders make decisions?

Strongly agree	Agree	Slightly agree	Neither agree nor disagree	Slightly disagree	Disagree	Strongly disagree

11. Accountability to your commander is the most important type of accountability?

Strongly agree	Agree	Slightly agree	Neither agree nor disagree	Slightly disagree	Disagree	Strongly disagree

12. A hostile environment will make you behave in a different way than a safe environment?

Strongly agree	Agree	Slightly agree	Neither agree nor disagree	Slightly disagree	Disagree	Strongly disagree

13. If ordered you would restrict a prisoners sleep to gain critical information?

Strongly agree	Agree	Slightly agree	Neither agree nor disagree	Slightly disagree	Disagree	Strongly disagree

14. It is acceptable to use nicknames for the enemy that you are fighting?

Strongly agree	Agree	Slightly agree	Neither agree nor disagree	Slightly disagree	Disagree	Strongly disagree

15. You should take action if you see poor behaviour by a soldier not in your platoon?

Strongly agree	Agree	Slightly agree	Neither agree nor disagree	Slightly disagree	Disagree	Strongly disagree

APPENDIX 2: ETHICAL DECISION MAKING COACHING TOOL

Coaches Guidance on Completion of the Ethical Decision-Making Coaching Tool

1. **Aim**. The aim of this document is to assist coaches with the completion of the Ethical Decision-Making Coaching Tool to support ethical development of trainees under pressure situations.
2. **Timeframe**. The process takes around half an hour. The Tool will require at least 10 minutes to conduct, the After Action Review with the trainee around 10 minutes and a final another 10 minutes for the trainee to complete their reflection and get a final debrief.
3. **Aim of the Tool**. The Tool should be used to coach trainees on any field training exercise to enhance both their ethical and unethical decision-making experiences. It is not for 'special ethical' serials but should be used when a member of staff identifies an action by any trainee that appears to be unethical.
4. **Introduction**. Firstly, ask the trainee to come up with the unethical/ethical action that you want to discuss. If they do not know what the coaching session is going to be about you need to guide them into the discussion subject.
5. **Role Model**. You are trying to get the trainee to identify an ethical role model and identify why they think they are a good role model for them (they will come back to this person at the end of the session).

6. **Situation Influencers**. You are trying to see if the trainee understands the power of the situation on them and what might have enhanced this power on them. You might not want to ask all the questions depending on the situation. Ideally, you would ask them about their emotions and level of fatigue.
7. **Common Behaviours**. You are trying to understand if the trainee has an appreciation for the effects on their cognitive ability or the ability of their team to act. You may not want to ask both questions depending on the situation.
8. **Accountability**. Three of the four types of accountability can be summarised by the Army Standards. You might want to ask only one of these questions to see if the trainee believed that their actions were lawful, appropriate or professional.
9. **Leadership**. This section is getting at the trainee's ability to consider either their leadership or their ability to respectfully challenge authority. You may want to ask the first question in all situations and then one of the others depending on the role of the trainee.
10. **Moral Compass**. This question is a simple one focused at the trainee's belief that they have exercised moral courage.
11. **Narrative**. Once the above questions have been answered, the trainee has 10 minutes to write a narrative in their own words and in the first person to summarise their experiences.
12. **Looking Back/Looking Forward**. The trainee will also be required to look back at their exemplar and write down how they think their role model would have acted. They are also asked to rate how confidently they believe they could use the STOP process in a future similar situation.
13. **Getting to 10**. When the trainee has competed their reflection, you can have a final chat with them. They will have rated their future confidence and your final chat should ask them what it would take to get them to a '10' in the future? A further question could be 'What would you do if faced with a similar situation again?' Why and How?

The trainee should be given the card so that they can reflect on it in their own time.

Appendix 2: Ethical Decision Making Coaching Tool

Ethical Decision-Making Coaching Tool				
Event:	Trainee:		Coach:	
	Introduction: We now have 10 mins to develop your Ethical Decision Making based on the S-CALM Model. What recent event would you like to talk about?			
	Role Model: Tell me who is your ethical role model? What is it that makes them different to others for you?			
	Situation Influencers: See Aide Memoire for details	Did you feel that you were operating in a 'Hostile Environment'? Do you believe that you were experiencing 'Normalised Violence'? Was there a 'Lack of Supervision or Leadership? Do you think you had enough time to prepare? During the action, what emotions do you think you were experiencing? How tired are you at present?		

The S-CALM Model

Common Behaviours: See Aide Memoire for details	Do you feel that during the action you experienced any of the individual common behaviours? Do you believe that your team displayed any of the group common behaviours?
Accountability:	Did you consider accountability before you acted? When you took your action, did you consider that it was lawful? When you took your action, did you consider that it was appropriate? When you took your action, did you consider that it was professional?

Appendix 2: Ethical Decision Making Coaching Tool

Leadership:	Was the commanders Intent clear to you and do you feel that you followed it? Do you think when making the decision that you applied the 'loneliness of command'? Was there an opportunity to display 'Intelligent Disobedience' during this action?
Moral Compass:	Do you believe that you did the morally correct thing?

Narrative: Considering the things just discussed, can you summarise the key moments as you experienced them?

Example: "I had about xx hours in the last xx days. The situation was (situational influencers) and I felt (Enhanced Emotions). I sensed that I/we acted (common behaviours). I considered that my actions were (lawful/appropriate/professional). I weighed up the situation and decided to (leadership action). I chose to (Moral Compass)."

Looking Back: What would your role model have done in this situation?

Looking Forward: How confident do you feel that you will be able to use the STOP process in a future similar situation?

1	2	3	4	5	6	7	8	9
No confidence								High confidence

APPENDIX 3: DECISION MAKING QUESTIONNAIRE 2

Please tick or circle the box which best reflects your answer to the 15 questions below:

Decision Questions

1. It is OK for a platoon commander to use the first names of their soldiers?

Strongly agree	Agree	Slightly agree	Neither agree nor disagree	Slightly disagree	Disagree	Strongly disagree

2. You can break the rules of war if you believe that they are not useful for your mission?

Strongly agree	Agree	Slightly agree	Neither agree nor disagree	Slightly disagree	Disagree	Strongly disagree

3. You should always obey an order by a senior commander?

Strongly agree	Agree	Slightly agree	Neither agree nor disagree	Slightly disagree	Disagree	Strongly disagree

4. A commander should never sleep when their soldiers are awake?

Strongly agree	Agree	Slightly agree	Neither agree nor disagree	Slightly disagree	Disagree	Strongly disagree

5. It is OK not to follow the Army's Values when you are not in uniform and off duty?

Strongly agree	Agree	Slightly agree	Neither agree nor disagree	Slightly disagree	Disagree	Strongly disagree

6. It is justifiable to kill a civilian in order to achieve the mission?

Strongly agree	Agree	Slightly agree	Neither agree nor disagree	Slightly disagree	Disagree	Strongly disagree

7. It is always best to go with your initial intuitive decision?

Strongly agree	Agree	Slightly agree	Neither agree nor disagree	Slightly disagree	Disagree	Strongly disagree

8. An enhanced emotional state leads to an increase in unethical behaviour?

Strongly agree	Agree	Slightly agree	Neither agree nor disagree	Slightly disagree	Disagree	Strongly disagree

9. It is justifiable to kill a civilian in order to save a soldier?

Strongly agree	Agree	Slightly agree	Neither agree nor disagree	Slightly disagree	Disagree	Strongly disagree

10. Fatigue has little affect the way that commanders make decisions?

Strongly agree	Agree	Slightly agree	Neither agree nor disagree	Slightly disagree	Disagree	Strongly disagree

11. Accountability to your commander is the most important type of accountability?

Strongly agree	Agree	Slightly agree	Neither agree nor disagree	Slightly disagree	Disagree	Strongly disagree

Appendix 3: Decision Making Questionnaire 2

12. A hostile environment will make you behave in a different way than a safe environment?

Strongly agree	Agree	Slightly agree	Neither agree nor disagree	Slightly disagree	Disagree	Strongly disagree

13. If ordered you would restrict a prisoners sleep to gain critical information?

Strongly agree	Agree	Slightly agree	Neither agree nor disagree	Slightly disagree	Disagree	Strongly disagree

14. It is acceptable to use nicknames for the enemy that you are fighting?

Strongly agree	Agree	Slightly agree	Neither agree nor disagree	Slightly disagree	Disagree	Strongly disagree

15. You should take action if you see poor behaviour by a soldier not in your platoon?

Strongly agree	Agree	Slightly agree	Neither agree nor disagree	Slightly disagree	Disagree	Strongly disagree

BIBLIOGRAPHY

Books

ADF. *Inspector-General of The Australian Defence Force Afghanistan Inquiry Report, Part 1 – The Inquiry, Part 3 – Operational, Organisation and Cultural Issues.* Canberra: ADF, 2020.

ADF. *Military Ethics.* Australia: Directorate of Information, 2021.

ADF. *Leadership.* Australia: Directorate of Information, 2021.

Allison, W. *My Lai: An American Atrocity in the Vietnam War.* Baltimore: Johns Hopkins University Press, 2012.

Arendt, H. *Eichmann in Jerusalem: A Report on the Banality of Evil.* New York: Viking Press, 1963.

Asch, S. 'Effects of Group Pressure upon the Modification and Distortion of Judgement'. *Group Leadership and Men,* edited by H. Guetzkow, Pittsburgh: Carnegie Press, 1951.

Bandura, A. 'Moral Disengagement'. *The Encyclopaedia of Peace Psychology, First Edition,* edited by Daniel Christie, New Jersey: Blackwell Publishing, 2012.

Bauman, Z. *Modernity and the Holocaust.* New York: Cornell University Press, 1989.

Bennett, H. *Fighting the Mau: The British Army and Counter-Insurgency in the Kenya Emergency.* Cambridge: Cambridge University Press, 2012.

Bercuson, D. *Significant Incident: Canada's Army, the Airborne, and the Murder in Somalia.* Toronto: McCleland & Stewart, 1996.

Blackman, A. *Marine A: My Toughest Battle.* London: Mirror Books, 2019.

Blake, N. *The Deepcut Review: A Review of the Circumstances Surrounding the Deaths of Four Soldiers at Princess Royal Barracks, Deepcut between 1995 and 2002.* London: The Stationery Office, 2006.

Boszko, J. *Encyclopaedia of the Holocaust Vol. 2.* London: Macmillan, 1990.

Breakwell, G. *The Psychology of Risk.* Cambridge: Cambridge University Press, 2007.

Bregmann, R. *Human Kind: A Hopeful History.* London: Bloomsbury, 2020.

Brown, M. and S. Seaton. *Christmas Truce.* London: Pan, 1994.

Browning, C. *Ordinary Men: Reserve Police Battalion 101 and the Final Solution in Poland.* New York: Harper Collins Books, 1998.

Bundeswehr, *Making the Right Decision – Acting Responsibly.* Bundeswehr, 2013.

Canada Commission of Inquiry into the Deployment of Canadian Forces to Somalia and Donna Winslow. *The Canadian Airborne Regiment in Somalia, a Socio-Cultural Inquiry: A Study Prepared for the Commission of Inquiry into the Deployment of Canadian Forces to Somalia*. Canadian Government Publishing, 1997.

Cencich, J. *The Devil's Garden: A War Crimes Investigators Story*. Nebraska: Potomac Books, 2013.

Cesarani, D. *Becoming Eichmann: Rethinking the Life, Crimes and Trial of a "Desk Murderer"*. Cambridge: Da Capo Dress, 2006.

Chaleff, I. *Intelligent Disobedience: Doing Right When What You're Told To Do Is Wrong*. Oakland: Berrett-Koehler Publishers, 2015.

Cialdini, R. *Influence: The Psychology of Persuasion*. New York: Harper Collins, 2007.

Clark, L. *Leadership Insight No.1 March 2017: The Intelligently Disobedient Soldier*. Camberley: The Centre for Army Leadership, 2017.

Coates, A. 'Culture, the Enemy and Moral Restraint in War'. *The Ethics of War: Shared Problems in Different Traditions*, edited by Richard Sorabji and David Rodin, London: Routledge, 2006.

Collett, N. *The Butcher of Amritsar: General Reginald Dyer*. London: Cambridge University Press, 2005.

Crompvoets, S. *Blood Lust, Trust and Blame*. Victoria: Monash University Publishing, 2021.

Davis, D. and V. Shackleton. *Psychology and Work*. London: Methuen, 1975.

Deakin, S. *Leadership: Proceedings of a Symposium Held at the Royal Military Academy Sandhurst, April 2014*. Sandhurst Occasional Paper No. 18. Sandhurst: Central Library, 2014.

Docherty, J. *Learning Lessons from Waco: When the Parties Bring their Gods to the Negotiation Table*. Syracuse: Syracuse University Press, 2001.

Draper, A. *Amritsar: The Massacre that ended the Raj*. London: Cassell, 1981.

Elkins, C. *Imperial Reckoning: The Untold Story of Britain's Gulag in Kenya*. Boston: Boston University African Studies Center, 2005.

Ellner, A. Paul Robinson and David Whetham. *When Soldiers Say No: Selective Conscientious Objection in the Modern Military*. Farnham: Ashgate, 2014.

Fairweather, J. *A War of Choice: Honour, Hubris and Sacrifice: The British in Iraq*. London: Vintage, 2012.

Festinger, L. *A Theory of Cognitive Dissonance*. Stanford: Stanford University Press, 1957.

Fredericks, J. *Black Hearts: One Platoon's Descent into Madness in Iraq's Triangle of Death*. London: Macmillan, 2010.

Gage, W. *The Report of the Baha Mousa Inquiry* London. The Stationery Office, 2011.

Gardner, H. *Leading Minds: An Anatomy of Leadership*. London: Harper Collins, 1996.

Goldhagen, D. *Hitler's Willing Executioners: Ordinary Germans and the Holocaust*. London: Abacus, 1997.

Gourevitch, P. and E. Morris. *Standard Operating Procedure: A War Story*. London: Picador, 2009.

Guroian, V. *Tending the Heart of Virtue*. Oxford: Oxford University Press, 1998.

Hagedorn, A. *The Invisible Soldiers: How America Outsourced Our Security*. New York: Simon & Schuster, 2014.

Hale, C. *Massacre in Malaya: Exposing Britain's My Lai*. Stroud: The History Press, 2013.

Hall, D., S. Pilbeam and M. Corbridge. *Contemporary Themes in Strategic People Management: A Case-based Approach*. Basingstoke: Palgrave Macmillan, 2013.

Hart Dyke, D. *Four Weeks in May: A Captain's Story of War at Sea*. London: Atlantic Books, 2008.

Hogg, M. and G. Vaughan. *Social Psychology, 7th Ed*. Harlow: Pearson, 2014.

Honig, J. W. and N. Both. *Srebrenica: Record of a War Crime*. London: Penguin, 1996.

ICRC. *The Roots of Behaviour in War: Understanding and Preventing IHL Violations*. Geneva: ICRC, 2004.

ICRC. *Humanity In Action – 2021*. Geneva: ICRC, 2021.

Jackson, M. *Solider: The Autobiography of General Sir Mike Jackson*. London, Batam Press, 2007.

Jamail, D. *Will to Resist: The Soldiers Who Refuse to Fight in Iraq and Afghanistan*. Chicago: Haymarket Books, 2011.

Janis, I. *Victims of Groupthink: A Psychological Study of Foreign Policy Decisions and Fiascos*. Boston: Houghton Mifflin, 1972.

Janis, I. *Groupthink: Psychological Studies of Policy Decisions and Fiascos*, 2nd ed. Boston: Houghton Mifflin, 1972.

Janis, I. *Groupthink*. Boston: Houghton Mifflin, 1982.

Jonas, G. *Vengeance*. London: Harper Perennial, 2006.

Kahneman, D. *Thinking, Fast and Slow*. New York: Farrar, Straus and Giroux, 2011.

Kahneman, D., O. Sibony and C. Sunstein. *Noise: A Flaw in Human Judgement*. London: William Collins Books, 2022.

Karssing, E. 'The E-Word (Emotions) in Military Ethics Education: Making Use of the Dual-Process Model of Moral Psychology' *Violence in Extreme Conditions: Ethical Challenges in Military Practice*, edited by E. Kramer and T. Molendijk, Cham: Springer, 2023.

Kaurin, P. *The Warrior, Military Ethics and Contemporary Warfare*. Farnham: Ashgate, 2014.

Kemp, S. *British Justice, War Crimes and Human Rights Violations*. Switzerland: Palgrave MacMillan, 2019.

Levi-Strauss, C. *Race and History*. Paris: UNESCO, 1952.

Lloyd, N. *The Amritsar Massacre: The Untold Story of One Fateful Day*. London: I.B. Tauris, 2011.

Mestrovic, S. *The Postemotional Bully*. SAGE Swifts, 2014.

Milgram, S. *Obedience to Authority: An Experimental View*. New York: Harper Collins, 1974.

Mileham, P. 'An Essay on Military Ethics' *Developing Leaders: A British Army Guide*, Camberly: The Royal Military Academy Sandhurst, 2014.

Miller, N., P. Matsangas and A. Kenney. 'The role of Sleep in the Military: Implications for Training and Operational effectiveness', *The Oxford Handbook of Military*

Psychology, edited by J. Laurence and M. Matthews, New York: Oxford University Press, 2012.

MOD. *Board of Inquiry – Report into the Loss of HMS Coventry*. London: MOD, 1982.

MOD. *The Government's Response to the Deepcut Review*. Crown Copyright, 2006.

MOD. *Values & Standards of The British Army*. London: HMSO, 2008.

MOD. *The Aitken Report: An Investigation into Cases of Deliberate Abuse and Unlawful Killing in Iraq in 2003 and 2004*. Crown Copyright, 2008.

MOD. *The Army Leadership Code: An Introductory Guide*. Camberley: The Centre for Army Leadership, 2015.

MOD. *Joint Doctrine Publication 04 Understanding and Decision-making*, 2nd ed. London: DCDC, 2016.

MOD. *The Good Operation: A handbook for those involved in operational policy and its implementation*. London: HMSO, 2017.

MOD. *Values & Standards of The British Army*. London: HMSO, 2018.

MOD. *A Practical guide to Military Ethics within the Submarine Service*. London: HMSO, 2019.

MOD. *Army Leadership Doctrine*. London: HMSO, 2021.

MOD. *Behavioural Science in Defence: A toolkit for practitioners*. London: HMSO, 2021.

MOD. *ADP Land Operations, Part 2: The Application of Land Power*. Warminster: Land Warfare Centre, 2022.

MOD. *Active Bystander Fundamentals*. MOD, 2023.

MOD. *A British Army Followership Doctrine Note*. Camberley: The Centre for Army Leadership, 2023.

Murray, D. *Bloody Sunday: Truths, Lies and the Saville Inquiry*. London: Biteback Publishing, 2012.

Murray, L. *Brains and Bullets: How Psychology Wins Wars*. London: Biteback, 2013.

NATO. *Leader Development for NATO Multinational Military Operations, STO-TR-HFM-286*. Boston Spa: NATO, 2022.

Northouse, P. *Leadership*, 9th ed. London: Sage, 2022.

Olusanya, O. *Emotions, Decision-making and Mass Atrocities*. Farnham: Ashgate, 2014.

Osiel, M. *Obeying Orders: Atrocity, Military Discipline and the Law of War*. London: Routledge, 2017.

Osland, J.S., et al. *Global Leadership. Research, Practice and Development*. Oxford: Routledge, 2008.

Perry, G. *Behind the Shock Machine: The Untold Story of the notorious Milgram Psychology Experiments*. New York: New Press, 2013.

Prescott, J., et al. *Ordinary Soldiers: A Study in Ethics, Law and Leadership* West Point: Center for Holocaust and Genocide Studies at West Point, 2014.

Razack, S. *Dark Threats and White Knights: The Somalia Affair, Peacekeeping, and the New Imperialism*. Toronto. University of Toronto Press, 2004.

Riley, J. *White Dragon: The Royal Welch Fusiliers in Bosnia*. Wrexham: The Royal Welch Fusiliers, 1995.

Robinson, P. 'Introduction: Ethics Education in the Military'. *Ethics Education in the Military*, edited by P. Robinson, N. de Lee, and D. Carrick, Aldershot: Ashgate, 2008.

Roeckelein, J. *Elsevier's Dictionary of Psychological Theories*. Fountain Hills: Elsevier, 2006.

Rohde, D. *Endgame: The Betrayal and Fall of Srebrenica, Europe's Worst Massacre Since World War II*. New York: Penguin Books, 2012.

Royal, B. *The Ethical Challenges of the Soldier: The French Experience*. Paris: Economica, 2012.

Saville, Lord. *Principal Conclusions and Overall Assessment of the Bloody Sunday Inquiry*. London: The Stationery Office, 2010.

Sherman, N. *The Untold War: Inside the Hearts, Minds, and Souls of Our Soldiers*. London: W.W. Norton, 2010.

Sibony, O. *You're About To Make A Terrible Mistake: How Biases Distort Decision-Making and What You Can Do to Fight them*. Croydon: Swift Press, 2021.

Skerker, M., D. Whetham and D. Carrick. *Military Virtues*. Havant: Howgate, 2019.

Stogdill, R. *Handbook of Leadership*. London: Collier Macmillan, 1974.

Talbert M. and J. Wolfendale. *War Crimes: Causes, Excuses and Blame*. Oxford: Oxford University Press, 2019.

Tripodi, P. 'Understanding Atrocities: What Commanders Can Do to Prevent Them'. *Ethics, Law and Military Operations*, edited by David Whetham, London: Palgrave, 2010.

The Centre for Army Leadership. *What Leaders Are Workshop*. Camberley: CAL, 2019.

US Army. *ADP 6-22: Army Leadership and The Profession*. Department of the Army, 2019.

US Army. *A Special Operations Forces Ethics Field Guide*. United States Special Operations Command, 2023.

Vincent, D. *Be, Know or do? An analysis of the Optimal Balance of the Be, Know, Do Leadership Framework in future Training at the Royal Military Academy Sandhurst*. Sandhurst Occasional Paper No. 20. Sandhurst: Central Library, 2015.

Vincent, D. *Ethical Warfare: Jus In Bello during the Initial Interventions into Iraq and Afghanistan*. Sandhurst Occasional Paper No. 25. Sandhurst: Central Library, 2019.

Vincent, D. 'Towards Jus Post Bellum: Ethical Warfare for stabilisation in Iraq and Afghanistan'. *Jus Post Bellum: Restraint, Stabilisation and Peace*, edited by Patrick Mileham, Leiden: Brill Nijhoff, 2020.

Vincent, D. and A. Muhl-Richardson. *An Analysis of the Effectiveness of the S-CALM Model of Ethical Leadership at the Royal Military Academy Sandhurst*. Sandhurst Occasional Paper No. 37. Sandhurst: Central Library, 2024.

Wagner, K. *Amritsar 1919: An Empire of Fear and Making of a Massacre*. London: Yale University Press, 2019.

Walker, M. *Why We Sleep*. London: Penguin, 2017.

Walzer, M. *Just and Unjust Wars*. New York: Basic Books, 1977.

Ward, I. and N. Miraflor. *Slaughter and Deception at Batang Kali*. Singapore: Media Masters, 2009.

Westley, R. and M. Ryan. *Operation Insanity: The Dramatic True Story of the Mission that Saved 10,000 Lives*. London: John Blake Publishing, 2016.
Whetham, D. 'Ethics Education and Training'. *Defence Academy Ethics Seminar: To Consider the Ethical Component of Military Capability*, edited by Patrick Mileham, London: Royal College of Defence Studies, 2011.
Williams, A. *A Very British Killing: The Death of Baha Mousa*. London: Vintage Books, 2013.
Woodward, S. with P. Robinson. *One Hundred Days: The Memoirs of the Falklands Battle Group Commander*. London: Harper Collins, 2003.
Yin, R. *Case Study Research: Design and Methods*, 2nd ed. London: Sage Publications, 1994.
Zimbardo, P. *The Lucifer Effect: How Good People Turn Evil*. Croydon: Rider Books, 2007.

Articles and Papers

Aloysius, S. 'The Role of Emotional Intelligence in Leadership Effectiveness', Conference Paper, Jaffna University Research Conference, October 2010.
Anderson, D. 'A Very British Massacre', *History Today, August 2006* (2006), 20-22.
Barnes, C., J. Schaubroeck, M. Huth and S. Ghumman. 'Lack Of Sleep And Unethical Conduct', *Organizational Behavior and Human Decision Processes*, Vol. 115 (2011), 169-180.
Bennett, H. 'The Baha Mousa Tragedy: British Army Detention and Interrogation from Iraq to Afghanistan', *British Journal of Politics and International Relations*, Vol. 16 (2014), 211–229.
Bhavnani, R. 'Ethnic Norms and Interethnic Violence: Accounting for Mass Participation in the Rwandan Genocide', *Journal of Peace Research*, Vol. 43 (2006), 651-669.
Blaskovich, J. 'Odyssey to Dante's Inferno', *Medicinski vjesnik*, Vol. 27, No. 1-2 (1995), 131-135.
Bradley, J. and S. Tymchuk. 'Assessing and Managing Ethical Risk in Defence', *Canadian Military Journal*, Vol. 13, No. 4, Autumn 2013 (2013), 6-16.
Brown, M. and L. Trevino. 'Ethical Leadership: A Review and Future Directions', *The Leadership Quarterly*, Vol. 17 (2006), 595-616.
Brown, M., L. Trevino and D. Harrison. 'Ethical Leadership: A Social Learning Perspective for Construct Development and Testing', *Organizational Behavior and Human Decision Processes*. Vol. 97 (2005), 117-134.
Brunner, J. 'Eichmann's Mind: Psychological, Philosophical, and Legal Perspectives', *Theoretical Inquiries in Law*, Vol. 1, No. 2 (2008), 1-35.
Bury, P. 'Maintaining Morality at the Tactical Level: A Junior Commander's Perspective', *British Army Review*, Vol. 150, Winter 2010/2011 (2011), 120-123.
Chang, Jenna. 'The Role of Anonymity in Deindividuated Behaviour: A Comparison of Deindividuation Theory and the Social Identity Model of Deindividuation Effects (SIDE)', *The Pulse*, Vol. 6, No. 1 (2008), 2-8.

Chin, Warren. 'Why Did It All Go Wrong? Reassessing British Counterinsurgency in Iraq', *Strategic Studies Quarterly, Winter 2008* (2008), 119-135.

Cushman, J. Chain of Command Performance of Duty, 2d Brigade Combat Team,101st Airborne Division, 2005-06. A Case Study Offered to the Center for Army Professional Ethic. 2011.

Cusumano, E. 'Diplomatic Security for Hire: The Causes and Implications of Outsourcing Embassy Protection', *The Hague Journal of Diplomacy*, Vol. 12 (2017), 27-55.

Darley, J. and B. Latane. 'Bystander Intervention in Emergencies: Diffusion of Responsibility', *Journal of Personality and Social Psychology*, Vol. 8, No. 4 (1968), 377–383.

Darley J. and D. Batson. 'From Jerusalem to Jericho: a Study of Situational and Dispositional Variables in helping Behavior', *The Research Experience, Experiment*, 1973, 191-214.

Dubnick, M. 'Accountability And Ethics: Reconsidering the Relationships', *International Journal of Organization Theory and Behavior*, Vol. 6, No. 3, Fall 2003 (2003), 405-441.

Duffy, A. 'Legacies of British Colonial Violence: Viewing Kenyan Detention Camps through the Hanslope Disclosure', *Law and History Review*, Vol. 33, No. 3 (2015), 489-542.

Duffy, M., T. McGirk, and B. Ghosh. 'The Ghosts of Haditha', *TIME*, Vol. 167, No. 24 (2006), 26-35.

Dutton, D., E. Boyanowsky and M. Harris Bond. 'Extreme mass homicide: From military massacre to genocide', *Aggression and Violent Behavior*, Vol. 10 (2005), 437–473.

Edmondson, A. 'Psychological Safety and Learning Behavior in Work Teams', *Administrative Science Quarterly*, Vol. 44, No. 2 (1999), 350-383.

Farnsworth, J., K. Drescher, J. Nieuwsma, R. Walser and J. Currier. 'The Role of Moral Emotions in Military Trauma: Implications for the Study and Treatment of Moral Injury', *Review of General Psychology*, Vol. 18, No. 4 (2014), 249 –262.

Faure, G. 'Negotiating with Terrorists: A Discrete form of Diplomacy', *The Hague Journal of diplomacy*, Vol. 3, No. 2 (2008), 179-200.

Festinger, L. 'A Theory of Social Comparison Processes', *Human Relations*, Vol. 7 (1954), 177-140.

Festinger, L. 'Cognitive Dissonance', *Scientific American*, Vol. 207, No. 4 (1962), 93-106.

Festinger, L., A. Pepitone and T. Newcomb. 'Some Consequences of Deindividuation in a Group', *Journal of Social Psychology*, Vol. 47 (1952), 382-389.

Flanagan, S. 'Losing Sleep', *Armed Forces Journal*, December 2011 (2011), 12-14.

Fredericks, J. 'The threat from within', *TIME Magazine*, Vol. 175, No. 7 (2010), 42-43.

Fromm, P., D. Pryer and K. Cutright. 'The Myths We Soldiers Tell Ourselves: and the Harm These Myths Do', *Military Review, September-October 2013* (2013), 57-68.

Gerring, J. 'What is a case study and what is it good for?', *American Political Science Review*, Vol. 98, No. 2 (2004), 341-354.

Glover, J. 'A Soldier and His Conscience', *Parameters*, Vol. 13, No. 1 (1983), 53-58.

Griggs, R. 'Coverage of the Stanford Prison Experiment in Introductory Psychology Textbooks', *Teaching of Psychology*, Vol. 42, No. 2 (2014), 195-203.

Hack, K. 'Devils that Suck the Blood of the Malayan People: The Case for Post-Revisionist Analysis of Counter-insurgency Violence', *War in History*, Vol 25, No. 2, April 2018 (2018), 202-226.

Hack, K. 'Everyone lived in Fear: Malaya and the British Way of Counter-insurgency', *Small Wars & Insurgencies*, Vol. 23, No. 4-5 (2012), 671-699.

Hegarty, H. and H. Sims. 'Some determinants of unethical decision behavior – An experiment', *Journal of Applied Psychology*, Vol. 63, No. 4 (1978), 451–457.

Hale, C. 'Batang Kali: Britain's My Lai', *History Today, July 2012* (2012), 3-4.

Hall, M. 'Why Leaders need a Morality Check', *United States Navel Institute*, Vol. 132, No. 4 (2006), 68-70.

Harden, M. 'Blackwater USA: The Success and Failures of the Worlds Most Powerful Mercenary Army in the War on Terror', *Pepperdine Policy Review*, Vol. 9 (2017), 23-45.

Hart, P. 'Preventing Groupthink Revisited: Evaluating and Reforming Groups in Government', *Organizational Behavior and Human Decision Processes*, Vol. 73, Nos. 2/3, February/March (1998), 306–326.

Haslam, N. 'Dehumanization: An Integrative Review', *Personality and Social Psychology Review*, Vol. 10, No. 3 (2006), 252–264.

Haslam S., A. and S. Reicher. 'Beyond the Banality of Evil: Three Dynamics of an Interactionist Social Psychology of Tyranny', *Personality and Social Psychology Bulletin*, Vol. 33 (2007), 615-622.

Isen, A. and P. Levin. 'Effect of feeling good on helping: Cookies and kindness,' *Journal of Personality and Social Psychology*, Vol. 21, No. 3 (1972), 384–388.

Janis, I. 'Groupthink', *IEEE Engineering Management Review*, Vol. 36, No. 1 (2008), 84-90.

Johnson, P. 'Effects of Groupthink on Tactical Decision-Making', Fort Leavenworth: USA Command & General Staff College, 2001.

Kahn, W. 'Psychological Conditions of Personal Engagement and Disengagement at Work', *Academy of Management Journal*, Vol. 33, No. 4 (1990), 692-724.

King, I. 'What is Ethical Leadership?', Royal College of Defence Studies, 2023.

Kleinman, C. 'Ethical Drift When Good People Do Bad Things'. *JONA's Healthcare Law, Ethics, and Regulation*, Vol. 8, No. 3 (2006), 72-76.

Krulak, C. 'The Strategic Corporal: Leadership in the Three Block War: Operation Absolute Agility', Center for Army Lessons Learned, Fort Leavenworth, 2002, 1-8.

Kweit, K. 'Hitler's willing executioners and "ordinary Germans" some comments on Goldhagen's ideas', Central European University, 1996.

Ilyas, S., G. Abid and F. Ashfaq. 'Ethical Leadership in sustainable Organizations: The Moderating Role of General Self efficacy and the Mediating Role of Organizational Trust', *Sustainable Production and Consumption*, Vol. 22 (2020), 195-204.

Johnson et al. 'Dress, Body and Self: Research in the Social Psychology of Dress', *Fashion and Textiles 2014*, Vol. 1, No. 20 (2014), 1-14.

Jones, T. 'Ethical Decision-Making by Individuals in Organizations – An Issue-Contingent Model', *Academy of Management Review*, Vol. 16 (1991), 366–395.

Kelman, H. 'Violence Without Moral Restraint: Reflections on the Dehumanization of Victims and Victimizers', *Journal of Social Issues*, Vol. 23 (1973), 251-261.

Larsen, R. 'Decision Making by Military Students Under Severe Stress', *Military Psychology*, Vol. 13, No.2 (2001), 89-98.

Linden, M., et al. 'A latent core of dark traits explains individual differences in peacekeepers' unethical attitudes and conduct', *Military Psychology*, Vol. 31, No. 6 (2019), 499-509.

Lindsay, D. 'Something Dark and Bloody', *The Quarterly Journal of Military History*, Vol. 25, No. 1, Autumn (2012), 50-59.

Lloyd, N. 'The Errors of Amritsar', *BBC History Magazine*, Vol. 10, No. 2, April 2009 (2009), 51-54.

Longenecker, C. and J. W. Shufelt. 'Conquering the Ethical Temptations of Command: Lessons from the Field Grades', *JPME Today, 2nd Quarter 2021* (2021), 36-44.

Luthans F. and B. J. Avolio. 'Authentic leadership development', *Positive organizational scholarship* Vol. 241, No. 258 (2003), 1-26.

Malamuth N. and S. Feshbach. 'Risky Shift in a Naturalistic Setting', *Journal of Personality* (1972), 38-49.

Mastroianni, G. 'The Person-Situation Debate: Implications for Military Leadership and Civilian-Military Relations', *Journal of Military Ethics*, Vol. 10, No. 1 (2011), 2-16.

Mastroianni, G. 'Looking Back: Understanding Abu Ghraib', *Parameters*, Vol. 43, No. 2, Summer 2013 (2013), 53-65.

McCormack, P. 'Preparing Professional Military Forces to Face Ethical Challenges in Future Military Operations', EuroISME Presentation (2015).

McCormack, P. 'Grounding British Army Values Upon an Ethical Good', British Army (unpublished) (2015).

McDermott, T. 'We Need to Talk About Marine A: Constant War, Diminished Responsibility and the Case of Alexander Blackman', ACSACS Occasional Paper No. 6 (2017).

Messervey, D., W. Dean, E. Nelson, and J. Peach. 'Making Moral Decisions Under Stress: A Revised Model for Defence', *Canadian Military Journal*, Vol. 21, No. 2, Spring 2021 (2021), 38-47.

Milgram, S. 'Behavioural Study of Obedience', *Journal of Abnormal and Social Psychology*, Vol. 67 (1961), 371-378.

Miller D., E. Smith and D. Mackie. 'Effects of Intergroup Contact and Political Predispositions on Prejudice: Role of Intergroup Emotions', *Group Processes & Intergroup Relations*, Vol. 7, No. 3 (2004), 221-237.

MOD. 'CLM Handbook' (2018).

Murphy, D. 'From My Lai to Abu Ghraib: The Moral Psychology of Atrocity', *Midwest Studies in Philosophy* (2007).

Mayer, D., M., K. Aquino, R. L. Greenbaum and M. Kuenzi. 'Who Displays Ethical Leadership, And Why Does It Matter? An Examination of Antecedents And

Consequences Of Ethical Leadership', *The Academy of Management Journal*, Vol. 55, No. 1 (February 2012) (2012), 151-171.

Napier, B. J. and G. R. Ferris. 'Distance in organizations', *Human Resource Management Review*, Vol. 3, No. 4 (1993), 321-357.

Navy Command. 'Telemeter-Internal Review', Navy Command Headquarters (2014).

Noor, J. 'Case Study: A Strategic Research Methodology', *American Journal of Applied Sciences*, Vol. 5, No. 11 (2008), 1602-1604.

Olsen O., S. Pallesen and J. Eid. 'The Impact of Partial Sleep Deprivation on Moral Reasoning in Military Officers', *SLEEP*, Vol. 33, No. 8 (2010), 1086-1090.

Olshfski, D. 'Accountability and Ethics in the Abu Ghraib Scandal', Paper delivered to the Ethics and Integrity of Governance Conference, Leuven, Belgium (2005), 78-95.

Olsthoorn, P. 'Situations and Dispositions: How to Rescue the Military Virtues from Social Psychology', *Journal of Military Ethics*, August (2017), 78-93.

Paul R. 'Hearts of Darkness: 'Perpetrator History' and why there is no why', *History of the Human Sciences*, Vol. 17 (2004), 211-251.

Porcelli, A. J. and M. R. Delgado. 'Acute Stress Modulates Risk Taking in Financial Decision Making', *Psychological Science*, Vol. 20, No. 3 (2009), 278–283.

Regan, M. and K. Mullaney. 'Emotion, Ethics, and Military Virtues', *Journal of Military Ethics*, Vol. 22, Nos. 3-4 (2023), 256-273.

Reinke, S. 'Service Before Self: Towards A Theory of Servant-Leadership', *Global Virtue Ethics Review*, Vol. 5, No. 3 (2004), 30-57.

Richardson, K. 'The Social Psychology of Evil: A Look at Abu Ghraib', Master's Thesis, Clemenson University (2012).

Robinson, K., B. McKenna and D. Rooney. 'The Relationship of Risk to Rules, Values, Virtues, and Moral Complexity: What We can Learn from the Moral Struggles of Military Leaders', *Journal of Business Ethics*, Vol. 179 (2022), 749–766.

Robinson, P. 'Ethics Training and Development in the Military', *Parameters*, Vol. 37, No. 1 (2007), 23-36.

Romzek, B. and M. Dubnick. 'Accountability in the Public Sector: Lessons from the Challenger Tragedy', *Public Administration Review*, May/June 1987 (1987), 228-238.

Royal Courts of Justice. 'Court of Appeal Judgement in the case between Regina and Alexander Wayne Blackman, 7 and 8 February 2017' (2017).

Samuelson, W. and R. Zeckhauser. 'Status Quo Bias in Decision Making', *Journal of Risk and Uncertainty*, Vol. 1 (1988), 7–59.

Scahill, J. 'Making a Killing', *Nation*, Vol. 285, No. 11 (2007), 21-24.

Schlesinger, J. 'Final Report of the Independent Panel to Review DoD Detention Operations', Independent Panel to Review DoD Detention Operations (2004).

Scullin, J. 'The Mau Mau Insurrection: The Failed Rebellion That Freed Kenya', Undergraduate Thesis, University of Washington Tacoma (2017).

Shalvi, S., O. Eldar and Y. Bereby-Meyer. 'Honesty Requires Time (and Lack of Justifications)', *Psychological Science*, Vol.23, No. 1264 (2012), 1264-1270.

Shay, J. 'Ethical Standing for Commander Self-Care: The Need for Sleep', *Parameters*, Vol. 28, No. 2 (1998), 93-105.

Shorey, G. 'Bystander Non-Intervention and the Somalia Incident', *Canadian Military Journal*, Winter 2000-2001 (2001), 19-28.

Short, A. 'The Malayan Emergency and the Batang Kalii Incident', *Asian Affairs*, Vol. 41, No. 3 (2010), 337-354.

Showry M. and K. V. L. Manasa. 'Self-Awareness-Key to Effective Leadership', *IUP Journal of Soft Skills*, Vol. 8, No. 1 (2014), 15-26.

Slim, W. 'Address to the Adelaide Division of the Australian Institute of management, 4 April 1957', *Australian Army Journal* (1957), 5-13.

Spranca, M., E. Minska and J. Baron, 'Omission and Commission in Judgment and Choice', *Journal of Experimental Social Psychology*, Vol. 27 (1991), 76-105.

Staszak, J. 'Other/Otherness', *International Encyclopaedia of Human Geography* (2008), 1-7.

Stanovich, K. and R. West. 'On the Relative Independence of Thinking Biases and Cognitive Ability', *Journal of Personality and Social Psychology*, Vol. 94, No. 4 (2008), 672-695.

Stockdale, J. 'Courage Under Fire: Testing Epictetus's Doctrines in a Laboratory of Human Behavior', Hoover Essays No. 6, Stanford University (1963).

Stockdale, J. 'The Stoic Warrior's Triad: Tranquillity, Fearlessness and Freedom', A lecture to the student body of The Marine Amphibious Warfare School, Quantico, Virginia (1995).

Stoner, J. 'A Comparison of Individual and Group Decisions Including Risk', Thesis, Massachusetts Institute of Technology, Boston (1961).

Stoner, J. 'Risky and Cautious Shifts in Group Decisions: The Influence of Widely Held Values', Working Paper, Massachusetts Institute of Technology, Boston (1967).

Suresh, A., S. Swarna Latha, P. Nair and N. Radhika. 'Prediction Of Fight Or Flight Response Using Artificial Neural Networks', *American Journal of Applied Sciences*, Vol. 11, No. 6 (2014), 912-920.

Thompson, H. 'Moral Courage In Combat: The My Lai Story', Lecture to Center for the Study of Professional Military Ethics (2003).

Thompson, M. and R. Jetly. 'Battlefield Ethics Training: Integrating Ethical Scenarios in high-Intensity Military Field Exercises', *European Journal of Psychotraumatology*, Vol. 5 (2014), 1-10.

Trevino, L., L. Hartman and M. Brown. 'Moral Person and Moral Manager: How Executives Develop a Reputation for Ethical Leadership', *California Management Review*, Vol. 42, No. 4 (2000), 128-142.

US Army. 'Article 15-6 Investigation of the 800th Military Police Brigade' (2004).

Weaver, G., L. Trevino and B. Agle. 'Somebody I Look Up To: Ethical Role Models in Organisations', *Organizational Dynamics*, Vol. 34, No. 4 (2005), 313-330.

Wolfendale, J. 'The Causes of War Crimes', *Journal of Military Ethics*, Vol. 22, Nos. 3-4 (2023), 274-288.

Zagzebski, L. 'Moral exemplars in theory and practice', *Theory and Research in Education*, Vol. 11, No. 2 (2013), 193–206.

Zilincik, S. 'The Role of Emotions in Military Strategy', *Texas National Security Review*, Vol. 5, No. 2 2022, 12-25.

Bibliography

Internet and Websites

Adair, J. "Action Centred Leadership, 2015." https://www.valuing-your-talent-framework.com/sites/default/files/resources/THK-032%20John%20Adair.pdf

APA. "APA Dictionary of Psychology." https://dictionary.apa.org

United States Holocaust Museum. "Auschwitz through the lens of the SS." http://www.ushmm.org/museum/exhibit/online/ssalbum/?content=2

"Auschwitz Scrapbook History." http://www.scrapbookpages.com/auschwitzscrapbook/History/Articles/HungarianJews.html

BBC. "Alexander Blackman: How should crimes on the battlefield be handled?" Hardtalk, BBC World Service, 22 January 2020. https://www.bbc.co.uk/sounds/play/w3csy9fj

BBC News. "Confrontation over Pristina Airport, Thursday 9 March 2000." http://news.bbc.co.uk/1/hi/world/europe/671495.stm

BBC News. "Fusiliers' battle to save Bosnians, 5 December 2002." http://news.bbc.co.uk/1/hi/wales/2535155.stm

BBC News. "Trial highlights camp's problems, 23 February 2005." http://news.bbc.co.uk/1/hi/uk/4287449.stm

BBC News. "Officer's orders 'led to abuse', 21 January 2005." http://news.bbc.co.uk/1/hi/uk/4193751.stm

BBC News. "Army medic Kylie Watson awarded Military Cross, 27 March 2011." https://www.bbc.co.uk/news/uk-12873091

BBC News. "Military Cross winner Kylie Watson 'shocked by award, 28 March 2011." https://www.bbc.co.uk/news/uk-northern-ireland-12886147

BBC News. "Royal Marines from 42 Commando return from Afghanistan, 25 October 2011." https://www.bbc.co.uk/news/uk-england-devon-15448969

BBC News. "Nato's crisis of trust in Afghanistan, 9 March 2012." https://www.bbc.co.uk/news/world-asia-17219153

BBC News. "Falklands War admiral Sandy Woodward dies aged 81, 5 August 2013." https://www.bbc.co.uk/news/uk-23575534

BBC News. "Colchester Army barracks 'sex videos' are disgraceful says minister. 9 June 2022." https://www.bbc.co.uk/news/uk-england-essex-61743173

BBC News. "Parachute Regiment Balkans deployment cancelled after sex videos. 18 June 2022." https://www.bbc.co.uk/news/uk-england-essex-61851456

Bentham, J. "A Fragment on Government, 1776." https://plato.stanford.edu/entries/bentham/#:~:text=Bentham%20launched%20his%20career%20as,A%20Comment%20on%20the%20Commentaries

Boyer, Peter. "A Different War, The New Yorker, 23 June 2002." https://www.newyorker.com/magazine/2002/07/01/a-different-war

Burke, Crispin. "No Time, Literally, For All Requirements." https://www.ausa.org/articles/no-time-literally-all-requirements

Bury, Paddy. "Pointing North – An Essay on Platoon Leadership", *Journal of Faith and War,* 2009. http://faithandwar.org/index.php/leadership/47-practice-and-mentoring-leadership/55-pointing-north-lessons-learned-by-an-infantry-platoon-commander-in-afghanistan

Cambridge Dictionary. https://dictionary.cambridge.org

Centre of Army Leadership. "Army Leadership Doctrine: What Leaders Are Workshop." https://www.army.mod.uk/who-we-are/our-schools-and-colleges/centre-for-army-leadership/

Fishback, Ian. "A Matter of Honor, Washington Post, Wednesday, September 28, 2005." https://web.archive.org/web/20200612233830/https://www.washingtonpost.com/wp-dyn/content/article/2005/09/27/AR2005092701527.html

Griffin, Ben. "Rogue heroes? 21 April 2023." https://www.forceswatch.net/comment/rogue-heroes/

Forbes. "How To De-Stress In 5 Minutes Or Less, According To A Navy SEAL." https://www.forbes.com/sites/nomanazish/2019/05/30/how-to-de-stress-in-5-minutes-or-less-according-to-a-navy-seal/?sh=7705ec7b3046

Forbes. "Quotes." https://www.forbes.com/quotes/author/thomas-b-macaulay/

Grayling, Professor A C. "Ethics versus morality, Warminster School, 2015." https://www.warminsterschool.org.uk/professor-a-c-grayling-lecture-ethics-versus-morality/

Hill Fernie, Anne. "Thinkers or Junkers? Germans in England 1860-1920 & Beyond, 2018." https://ragdeduniversity.co.uk/2018/09/15/germans-in-england-1860-1920/

Kennedy, John F. "Profile in Courage Award on May 16, 2005." https://www.jfklibrary.org/events-and-awards/profile-in-courage-award/award-recipients/joseph-darby-2005

Kant, Immanual. "Groundwork of the Metaphysic of Morals, 1785." https://human.libretexts.org/Bookshelves/Philosophy/Ethics_(Fisher_and_Dimmock)/2%3A_Kantian_ethics

King, Anthony. "The Very British Affair of Marine A, War on the Rocks, 2016." https://warontherocks.com/2017/04/the-very-british-affair-of-marine-a/

King's College London, Centre for Military Ethics. "Armouring Against Atrocity, 2016." https://militaryethics.uk/en/course/library

Klein, Gary. "Performing a Project Premortem", Havard Business Review, September 2007." https://hbr.org/2007/09/performing-a-project-premortem

Independent. "Courageous Army medic awarded Military Cross, 27 March 2011." https://www.independent.co.uk/news/uk/home-news/courageous-army-medic-awarded-military-cross-2254172.html

Lanchester Telegraph. "Baha Mousa inquiry: QLR soldiers were left to maintain order, 8 September 2011." https://www.lancashiretelegraph.co.uk/news/9240412.baha-mousa-inquiry-qlr-soldiers-left-maintain-order/

Morrison, David. "The standard you walk past is the standard you accept, ADF investigation, 2013." https://speakola.com/ideas/david-morrison-adf-investigation-2013

Bibliography

Mullin, Shanelle. "Is social proof really that important? Here's how to use it, 30 August 2023." https://cxl.com/blog/is-social-proof-really-that-important/#h-what-is-social-proof

National Heart, Blood and Lung Institute. "Sleep Brochure, 2018." https://www.nhlbi.nih.gov/resources/sleep-brochure

National Research Council. "Human Behavior in Military Contexts, The National Academies Press, 2008." https://doi.org/10.17226/12023

Newristics. "Authority Bias, Heuristic Encyclopaedia." https://newristics.com/heuristics-biases/authority-bias

New York Times. "Blackwater Shooting Scene Was Chaotic, 10 April 2009." https://web.archive.org/web/20090410140405/http://www.nytimes.com/2007/09/28/world/middleeast/28blackwater.html

New York Times. "3 Blackwater Guards Called Baghdad Shootings Unjustified, 16 January 2010." https://web.archive.org/web/20170227170643/http://www.nytimes.com/2010/01/17/world/middleeast/17blackwater.html

Parliamentary Archives. "British Parliamentary Papers, House of Commons Debate: Hola Detention Camp, Volume 607, 16 June 1959." https://hansard.parliament.uk/Commons/1959-06-16/debates/b8d721fb-856c-4c71-8ecf-b05f477837dd/HolaDetentionCamp

Parliamentary Archives. "British Parliamentary Papers, House of Commons Debate: Bosnia, Volume 260, 31 May 1995." https://hansard.parliament.uk/Commons/1995-05-31/debates/526f47de-0bc5-4417-90f7-a73fdcb93a65/Bosnia?highlight=gorazde#contribution-80e4115b-2f7a-40e4-83ae-6f3bac0619c6

Parliamentary Archives. "British Parliamentary Papers, House of Commons Debate: Defence, Volume 264, 17 October 1995." https://hansard.parliament.uk/Commons/1995-10-17/debates/9fa3f440-5838-4818-b177-e05b13e768db/Defence?highlight=royal%20welch%20fusiliers#contribution-4484f852-2f87-4187-a28d-d95966ad31c2

Raffaele, Danielle. "How could 'ordinary men' become genocidal killers in the Holocaust? 1 January 2010." https://core.ac.uk/display/41235087?utm_source=pdf&utm_medium=banner&utm_campaign=pdf-decoration-v1

Raimondo, Tony. "The My Lai Massacre: A Case Study, School of the Americas, Fort Benning." https://downloads.paperlessarchives.com/p/Eyjx/

Sean Rayment. "SAS soldier refuses to fight in Iraq, *The Age*, 13 March 2006." https://www.theage.com.au/world/sas-soldier-refuses-to-fight-in-iraq-20060313-ge1x6i.html

Sky News. "Royal Marine Guilty Of Murdering Afghan Fighter, Friday 8 November 2013." https://news.sky.com/story/royal-marine-guilty-of-murdering-afghan-fighter-10428900

Slim, W, 'Address by Field Marshal The Viscount Slim On 14 October 1952 toOfficer Cadets of The Royal Military Academy Sandhurst', https://www.pnbhs.school.nz/wp-content/uploads/2015/11/Slim.pdf

Quote Investigator. "The Only Thing Necessary for the Triumph of Evil is that Good Men Do Nothing. 2010." https://quoteinvestigator.com/2010/12/04/good-men-do/

Smith, Bryan. "Witness at Haditha, *Chicago Magazine July 2008*, 2008." https://www.chicagomag.com/Chicago-Magazine/July-2008/Witness-at-Haditha/

Stanić, Miloš. "27 Years On, Croatia Tries Serbs for Vocin Massacre, 2018." https://balkaninsight.com/2018/10/15/27-years-on-croatia-tries-serbs-for-vocin-massacre-10-10-2018/

The Falklands War. "HMS Coventry D118." https://www.hmscoventry.co.uk/d118/falklands/

The Guardian. "I'm not going to start Third World War for you Jackson told Clark, 2 August 1999." https://www.theguardian.com/world/1999/aug/02/balkans3

The Guardian. "Alexander Blackman's company was out of control, claims former comrade, 15 March 2017." https://www.theguardian.com/uk-news/2017/mar/15/alexander-blackmans-company-was-out-of-control-claims-former-comrade

The Guardian. "Correct a black mark in US history: former prisoners of Abu Ghraib get day in court, 24 April 2024." https://www.theguardian.com/world/ng-interactive/2024/apr/14/abu-ghraib-iraq-torture-abuse.

The Independent. "Army cancels Balkans deployment after paratroopers orgy video, 18 June 2022." https://www.independent.co.uk/news/uk/army-orgy-video-kosovo-bosnia-b2104022.html

The Long, Long Trail. "Life in the trenches of the First World War." https://www.longlongtrail.co.uk/soldiers/a-soldiers-life-1914-1918/life-in-the-trenches-of-the-first-world-war/

The Royal Welch Fusiliers Museum Archive. "Goražde Force." https://www.rwfmuseum.org.uk/archives.html

The Telegraph. "David Hart Dyke, the captain of HMS Coventry recalls the horror of his ship sinking in the Falklands War: 'It was black, with people on fire', 18 May 2012." https://www.telegraph.co.uk/news/worldnews/southamerica/falklandislands/9272406/David-Hart-Dyke-the-captain-of-HMS-Coventry-recalls-the-horror-of-his-ship-sinking-in-the-Falklands-War-It-was-black-with-people-on-fire....html

Thomas, Rob. "Acts of Omission vs. Commission, 2018." https://robdthomas.medium.com/acts-of-omission-vs-commission-4d494a6b0ec8

SSAFA – Falklands 40. "David Hart Dyke." https://www.ssafa.org.uk/support-us/our-national-campaigns/falklands-40/falklands-40-david-hart-dyke

University of Leicester. "Organisational Psychology." https://www.le.ac.uk/oerresources/psychology/organising/page_16.htm

UN. "Agreement for the prosecution and punishment of the major war criminals of the European Axis, Signed in London, on 8 August 1945, 1951." https://www.un.org/en/genocideprevention/documents/atrocity-crimes/Doc.2_Charter%20of%20IMT%201945.pdf

US Navy. "Combat Tactical Breathing." https://www.med.navy.mil/Portals/62/Documents/NMFA/NMCPHC/root/Documents/health-promotion-wellness/psychological-emotional-wellbeing/Combat-Tactical-Breathing.pdf

WCASA. "Social Norms toolkit: The Normalization of Violence." https://www.wcasa.org/wp-content/uploads/2020/01/PDFforToolkitNormalizationofViolence.pdf

Warrant Officer Historical Foundation. "The Forgotten Hero of My Lai: The Hugh Thompson Story." https://warrantofficerhistory.org/PDF/Forgotten_Hero_of_My_Lai-WO_Hugh_Thompson.pdf

Welch, Suzy. "The Rule of 10-10-10." https://www.oprah.com/spirit/suzy-welchs-rule-of-10-10-10-decision-making-guide/all

Interviews

Richard Westley, Interview with author, 2 October 2023.

INDEX

Abu Ghraib Abuse 66, 67, 107, 120, 121, 148, 149, 151, 157, 164
Accountability 6, 7, 15, 37, 52, 101, 104, 107, 123, 136, **137-152**, 155, 185, 186, 187, 190, 197, 200, 201, 207
Allied Rapid Reaction Corp 110
Aloysius, Shanthakumary 99, 107
Amritsar Massacre 14, 37 42-43,
Andreotta, Specialist Glenn 114, 115, 182, 183
Arendt, Hannah 9, 13, 44, 60
Asch, Solomon 38,
Authority Bias 14, 15, 55, **60-61**, 80, 83, **127-128**, 136
Australian Defence Force 5, 11, 15, 18, 32, 36, 39, 61, 63, 73, 86, 103, 130, 139, 156, 159, 160, 161, 164, 166, 172, 179, 188
Australian SAS 39, 47

Baha Mousa 14, 20-22
Banality of Evil **9**, 13, 44, 104
Banality of Good 104, 182
Bandura, Albert 22, 46, 49, 66, 68
Batang Kali Massacre 14, 15, 37, 47, 106, 164
Batson, Daniel 46, 47
Blackman, Sergeant Alexander 5, 15, 74-84, 165,
Blackwater Security 57, 58, 126
Blake, QC, Nicholas 53,

Bloody Sunday 14, **23**
Brereton Report 39
British Army's Core Values (*see also* Values and Standards) 4, 5, 171, **173-178**
Browning, Christopher 60, 117
Bundeswehr 177
Burke, Edmund 112, 113,
Bush, President George W. 30, 54, 68, 151
Bystander Effect 14, 15, 37, **49-50**, 82, 83, **112-114**,

Calley, Lieutenant William 40, 115, 156, 164
Camp Breadbasket Abuse 14, 26-27
Challenge 25, 82, 91, 104, 105, 109, 115, 116, 122, 123, 124, 154, 156, **163-168**, 200,
Chen, Private Danny 64
Christmas Day Truce 15, 132-133
Chuka Massacre 14, 55, 61-62
Cialdini, Robert 10, 104, 127
Clark, General Wesley 110, 111
Clark, Professor Lloyd 109
Cognitive Dissonance Theory 14, 15, 37, **46-48**, 80, 81, 83, 111, 119, 136, 190
Colburn, Specialist Lawrence 114, 115, 182, 183
Combat Tactical Breathing 88, 189

Index

Communication and Applied Behavioural Science (CABS) 2, 6, 192
Consequentialism 170, 171, 179
Conformity 14, 15, 36, **37-41**, 58, 59, 82, **104-106**, 136, 141
Coventry, HMS 15, 105-106
Cowan, John 45
Cunningham, General Sir Alan 130

Darby, Sergeant Joseph 15, 16, 107-108, 121, 148, 152, 163
Darley, John 46, 47, 49
Deepcut Barracks Review 14, 37, 52-53
Dehumanisation 14, 15, 55, 63, 64, **65-67**, 68, 78, 81, 83, 129, 130, 132, 136
Demonisation 14, 15, 55, **68-69**, 130, 132, 134, 136
Deindividuation 15, 37, **41-43**, 67, 107, 108,
Deontology 170, 171, 179
Dien Bien Phu 15, 111,
Diffusion of Responsibility 49, 51, 58, 83, 107, 112, 115, 141,
Dyer, Brigadier-General Reginald 42-43

Eichmann, Adolf 9, 10, 60,
England, Private Lynndie 67, 107, 108, 120, 151
Ethical Decision Making Coaching Tool ix, x, 16, 181, 186, 191, **199-202**
Ethical Hierarchy 186, **193-194**
Ethical Leadership RMAS Research Survey 5, 73, 88, 94, 98, 109, 113, 127, 129, 140, 144, 156, 175, 180, 186, 191

Falklands War 15, 105, 128
Fatigue 7, 13, 18, 19, **27-31**, 32, 34, 72, 79, 86, 94-96, 117, 197, 200, 207
Faure, Guy 68
Festinger, Leon 37, 41, 46, 104,
Fern, Guardsman James 106, 107
Franceschi, Sergeant Paul 15, 111, 112
Froissard, Private 1st Class 112

Graner, Corporal Charles 107, 108, 120-121
Geneva Conventions 21, 81, 143, 150, 151, 193
Griffin, Trooper Ben 15, 124
Groupthink 15, 36, 37, 54, **55-57**, 58, 63, 81, 104, 112, 122, 123-124, 125, 141,

Hart Dyke, Captain David 105-106, 156, 161
Hierarchical Accountability **140-141**, 144, 145, 148, 149, 152
Hola Camp Massacre 14, 37, 45
Hostile Environment 13, 18, 19, **20-22**, 34, 40, 72, 75, 76, 83, 86, 87-90, 93, 190, 196, 201, 208
Hughes, Sergeant 106
Humphreys, Colour Sergeant Pete 92-93, 101

Intelligent disobedience 109, 111, 128, 165, 166, 170, 201
International Committee of the Red Cross (ICRC) 11, 38, 56, 61, 66, 68,
International Humanitarian Law 81, 121, 141, 142, 145, 150, 151, 171

Jackson, Lieutenant General Mike 15, 110-111
Janis, Irving 55-56, 123
Just War tradition 171, 185, 190,

Kahneman, Daniel 35, 72, 188, 189
Kaplinski, Brigadier General Janis 120, 149, 150
Karremans, Lieutenant Colonel Thom 26
Kelman, Herbert 65
King's College London 119, 120
Kohlberg, Lawrence 118, 119, 120

Laissez-faire Leadership 24, 93, 148
Latane, Bibb 49
Law of Armed Conflict 6, 114, 171
Legal Accountability 141, 142, 143, 145, 150, 151
Lee, Lieutenant Colonel Oliver 76, 77, 80
Levi-Strauss, Claude 63
Loneliness of Command 94, 113, 153, 154, 155, 156, 157, 168, 187, 190, 194, 201

Mau Mau 45, 61, 62
Milgram, Dr Stanley 10, 11, 13, 43, 44, 60, 117, 127
Moral Compass 6, 7, 16, 101, 104, 111, 123, 132, 136, 145, **170-184**, 185, 186, 187, 190, 194, 200, 201, 202
Morrison, Lieutenant General David 113
My Lai Massacre 14, 15, 16, 37, 38, 40, 64, 80, 114, 115, 120, 157, 163, 164, 182-184

Newcomb, Theodore 41
Nisour Square Massacre 55, 57, 58, 126
Normalised Violence 13, 14, 18, 19, 20, 22, 23, 34, 76, 90, 91, 92, 201
North Atlantic Treaty Organisation (NATO) 5, 13, 27, 110, 130, 131, 169
Nuremberg Trials 108

Obedience 14, 15, 37, **43-45**, 60, 61, 62, 108-111, 127, 136, 141, 177, 192,
Ordiway, Major Ben x, 191
Othering 14, 15, 55, **63-65**, 104, 129-131, 136, 137, 192

Parachute Regiment, 3rd Battalion 16, 23-24
Parachute Regiment, Canadian 50
Pepitone, Albert 41
Personality 12, **117-120**, 159,
Phillabaum, Lieutenant Colonel Jerry 149
Police Battalion 101 14, 15, 55, 59, 116
Political Accountability 141, 143, 148, 151
Power of the Situation ix, v, 9, **11-13**, 14, 15, 18, 34, 36, 37, 39, 44, 48, 50, 54, 71-72, 73, 79, 81, 82, 83, 84, 102, 117, 118, 120, 173, 176, 179, 182, 186, 187, 190, 193, 194, 200
Pre-mortem 187
Pristina Airport 15, 110
Professional Accountability 138, 142, 149, 152
Psychological Safety 111, 136, 165, 168

Index

Reflection 86, 187, 188, 191, 199, 200
Reinke, Sandra 159
Remedios, Guardsman Victor 106, 107
Resource, Lack of 13, 14, 18, 19, 27, 29, 72, 78, 83, 86, 94, 96, 97, 186,
Risky Shift 14, 15, 37, 54, 55, 58, 59, 60, 72, 80, 81, 83, 125, 126, 136
Robinson, Paul 176-177
Royal Military Academy Sandhurst (RMAS) vii, x, 2, 5, 6, 16, 35, 72, 73, 88, 94, 98, 109, 113, 122, 127, 129, 140, 144, 155, 156, 174, 175, 180, 181, 182, 186, 191, 192
Royal Welch Fusiliers, 1st Battalion 15, 87, 89, 96, 97, 100, 101, 187

Samuelson, William 51
Sanders, General Sir Patrick 16, 154, 168-169
Self-Awareness 4, 99, 107, **187-188**,
Shinseki, General Eric 115
Sibille, First Lieutenant Josef 15, 134-135
Slim, Field Marshal Sir William 155, 161
Situational Influencers viii, 7, 13, 15, **18-19**, 25, 31, 34, 37, 38, 70, 72, 74, 75, 80, 81, 83, 85-87, 101, 102, 103, 135, 186, 187, 190, 192, 194, 202
Sleep **28-29**, 66, 79, 85, 94, 95, 96, 109, 134, 196, 198, 206, 208
Smith, Lieutenant General Rupert 101
Social Comparison Theory 14, 15, 36, **37-39**, 58, 104-106, 136
Somalia Affair 14, 37, 50, 143
Srebrenica 14, 25-26, 87, 101

Status Quo Bias 14, 15, 35, 37, **51-53**, 54, 115-117, 136
Stockdale, Admiral James 99
Stoner, James 58, 125
STOP Protocol ix, 7, 16, 105, 186, 187, 188, **189-190**, 194, 200, 202
Stressors 12, 32, 78
Supervision, Lack of 13, 14, 18, 19, **24-27**, 39, 72, 77-78, 83, 86, 93, 141, 157, 201
System 1 Thinking 32, 35-36, 72-73, 83, 189
System 2 Thinking 73, 84, 88, 187-189

Taguba, Major General Antonio 148-150
Taliban 5, 68, 74, 76, 80, 81, 130, 137
Thompson, Warrant Officer Hugh 15, 16, 89, 114-115, 120, 163, 171, 182-184

Utilitarianism 170-171
US Army 5, 14, 15, 27, 28, 30, 36, 40, 54, 64, 65, 73, 86, 99, 103, 108, 115, 121, 137, 138, 149, 158, 159, 160, 162, 164, 166, 172, 176, 177, 179, 184, 185, 188, 191

Values and Standards (*see also* British Army's Core Values) 5, 6, 84, 172, 176, 179, 192, 193, 194
Vietminh 112
Virtue Ethics 3, 5, 16, 170-171, 172-173, 175-177, 179, 194
Voćin Massacre 14, 55, 69

Watson, Lance Corporal Kylie 15, 130-131
Watson, Corporal Christopher 82, 83

West Belfast 86, 96
Weak Leadership 13, 14, 17, 18, 19,
    24-26, 72, 78, 83, 93, 101
Westley, Major Richard x, 15, 87,
    89-93, 96-101, 163,
Wolfendale, Professor Jessica 117
Woodward, Admiral Sandy 15,
    128-129

Zeckhauser, Richard 51
Zimbardo, Dr Phillip 10, 12, 13